Socio-Economic Impacts of Landmines in Southern Kurdistan

Socio-Economic Impacts of Landmines in Southern Kurdistan

By

Almas Heshmati and Nabaz T. Khayyat

CAMBRIDGE
SCHOLARS
PUBLISHING

Socio-Economic Impacts of Landmines in Southern Kurdistan,
by Almas Heshmati and Nabaz T. Khayyat

This book first published 2013

Cambridge Scholars Publishing

12 Back Chapman Street, Newcastle upon Tyne, NE6 2XX, UK

British Library Cataloguing in Publication Data
A catalogue record for this book is available from the British Library

ISBN (10): 1-4438-4198-6, ISBN (13): 978-1-4438-4198-6

To the Victims of Landmines and Unexploded Ordnances
in Southern Kurdistan.

To our beloved:
Emma-Shirin, Sara-Shilan and Lisa-Sharmin Heshmati,
and Awaz, Hamma, Naz and Baz Khayyat.

TABLE OF CONTENTS

LIST OF ABBREVIATIONS

ADO	Aras Demining Organization
AP	Anti-personnel mines
AT	Anti-tank mines
AV	Anti-vehicle mines
BIC	Bayesian Information Criterion
CBMRE	Community Based Mine Risk Education
CMA	Cranfield Mine Action
DMA	Directorate of Mine Action
ENRP	Electrical Network Rehabilitation Project
EOD	Landmines and Unexploded Ordnance Disposal
ERW	Explosive Remnants of War
EU	European Union
GDMA	General Directorate for Mine Action
GDSV	Governorate, District, Sub District, and Villages
GIC	Geneva International Centre for Humanitarian Demining
GICHD	Geneva International Center for Humanitarian Demining
GIS	Geographic Information System
GOI	Government of Iraq
HI	Handicap International
HMA	Humanitarian Mine Action
HO	Head Office
ICBL	International Campaign to Ban Landmines
ICRC	international Committee of Red Cross
IDP	Internally Displaced Person
IKMAA	Iraqi Kurdistan Mine Action Authority
IKMAC	Iraqi Kurdistan Mine Action Centre
ILIS	Iraq's Landmine Impact Survey
IMAS	International Mine Action Standards
iMMAP	Information Management Mine Action Programs
IMSMA	Information Management System for Mine Action
KDP	Kurdistan Democratic Party
KRG	Kurdistan Regional Government
LIS	Landmine Impact Survey
MCO	Mine Coordination Office
ME	Ministry of Environment

MA	Mine Awareness
MAG	Mines Advisory Group
MAG-DB	Mine Advisory Group Database
MAIS	Mine Action Information System
MAP	Mine Action Programme
MCT	Manual Clearance Teams
MDD	Mine Detection Dogs
MENA	Middle East and North Africa
MoPDC	Ministry of Planning and Direct Cooperation
MRE	Mine Risk Education
MVA	Mine Victim Assistance
NGO	Non-Governmental Organization
NMAA	National Mine Action Authority
NPA	Norwegian People's Aid
NSAG	Non-State Armed Group
OASIS	Operational Activity Security Information System
OCHA	Office for the Coordination of Humanitarian Affairs
OIP	UN Office of the Iraq Program (OIP
OPS	Operations
PCIA	Post clearance impact assessment
PDO	Pirmam Demining Organization
PIA	Post Impact Assessment
PLANS-DB	Plans database
PUK	Kurdistan Patriotic Union
QA	Quality Assurance
QAM	Quality Assurance Monitor
QC	Quality Control
RMAC	Regional Mine Action Centers
SL1	Survey Level One
STATA	Statistical Software package
SUMMIT	Spatial Management Information Tool
SWG	Survey Working Group
TDO	Tiroj Demining Organization
TNT	Trinitrotoluene
TV	Television
UN	United Nations
UNAMI	UN Assistance Mission for Iraq
UNDP	United National Development Program
UNHCR	UN High Commissioner for Refugees
UNICEF	United Nation Children's Fund
UNIDR	United Nations Institute for Disarmament Research

UNMAS	UN mine coordination offices
UNOIP	United Nations Office of the Iraq Program
UNOPS	UN Office for Project Services
UNOPS-MAP	United Nations Office for Project Services Mine Action Program
UNOPS-MAIS	UNOPS Mine Action Information System
USA	United States of America
UXO	Unexploded ordnance
VA	Victim Assistance
VIF	Variance Inflation Factors
VVAF	Vietnam Veterans of America Foundation
WHO	World Health Organization

PREFACE

The first use of landmines was in the middle of the eighteenth century, initially in the form of anti-tanks to subdue this new invention. In order to prevent cleaning the fields of anti-tanks, anti-personnel mines were developed and sown throughout anti-tank minefields to deter enemy soldiers from entering. Landmines are usually very simple devices in terms of technology and readily manufactured anywhere. Although mines were made mainly from metal, plastic and ceramic, wooden landmines were also designed to make their detection difficult. The unique characteristic of a mine is in its time-delayed explosive function, allowing it to remain inactive until hit by a victim. This unique and inexpensive property has led mines to be used intensively during conflicts to block enemies and opposition forces from accessing large pieces of land. Landmines are victim-activated and indiscriminate in their targets and impact—whoever triggers a mine, whether an animal, a child, a farmer, or a soldier, becomes its victim.

Worldwide landmine contamination is considered a serious and hazardous issue to humankind. There is an estimated 100 million anti-personnel mines underground and another 100 million stockpiled around the globe. Evidence shows that if the use of mines were to be stopped immediately, it would take another one hundred years and an enormous cost to clear the existing landmines in use. The existence of mines is considered a vital socio-economic and environmental problem facing many countries exposed to their use. Despite significant action, landmines still cause displacement, significant threats and barriers to economic and social development in many parts of the world, and severe poverty and underdevelopment to the affected regions. The additional significant medical, psychosocial, political and economic impacts of landmines are many. Mines prevent the return of normal and productive life to rural areas and maintain economic, social and political pressures on urban life.

Landmines have strong negative effects on the process of societies' reintegration following a situation of political instability. The Kurdistan region, for example, was subject to those phenomena due to continuous conflict and wars, with the civilians forced into dangerous mined areas in order to access basic resources to satisfy their daily lives. Injuries caused by mines affect the life of survivors and their families as well as the medical infrastructure of the mine-affected community. The concept of human rights covers mostly the political, civil and economic aspects while the concept of human security focuses on the viewpoints of individuals to protect them from threats to their lives, livelihoods and dignity and to utilize their full potentials. The economic impact of landmines can be seen as one of the determinant factors of economic security as it prevents people in the affected areas from working, and victims with disabilities face difficulties in finding employment and remain dependent.

Several security aspects are vital for humanity. For example, food security ensures that individuals have access to physical and basic economic foodstuffs. Health security ensures a minimum level of providing basic health services and protection. Environmental security aims to protect people from the threats of nature. Personal security aims to protect people from physical violence and crime. Community security aims at protecting people from loss of traditional values and relations of violence. In general, threats and injuries from landmines drastically impact the wellbeing of individual and societies.

The Kurdistan region of Iraq is fatally contaminated by mines and unexploded ordinances (UXOs) due to different armed conflicts in different time periods. Landmines and UXOs have caused more than 12,800 victims in the region since the early 1970s, of which more than 10,500 were killed. Landmines constitute a substantial barrier to rehabilitation and development activities in the region. The impacts of landmine contamination and related accidents can be categorized in terms of its socio-economic effects in three categories: impact on human life, blockage of access to necessary resources and the distortion of human attitude for various reasons. The outcome of mine action in Kurdistan can be improved by using more resources, better techniques and taking advantage of experience locally and internationally. Mine action refers to activities aimed at reducing the social, economic and environmental impacts of landmines. The objective is to reduce the landmine risk to the extent that people can live safely, free from constraints.

Resources refer to the mine action organizations, capacity, responsibility and resource availability. After political agreement on the federal system of Iraq, the responsibility of mine action was transferred from international organizations and local agencies to the central mine action agency. From the Kurds perspective, the transfer of responsibility to a central mine action agency to plan and implement demining is a major step backwards. This is evidence of a lack of awareness of the regional government to the mine issue. This action is not appropriate considering the Iraqi government's crime against the Kurds, including mass destruction and systematic minelaying in the Kurdistan region. The collected information might be misused by the Iraqi army and security forces in future conflicts. The regional government should actively promote local capacity and force the central government to allocate a special budget to clean up their remains and to rehabilitate the victims of their crime.

Two important principles of the international humanitarian law related to the use of certain conventional weapons like landmines are military necessity and the proportionality and adaptation of the use of such weapons. The above principles refer only to the use of landmines but not to production and supply. It is clear that the international humanitarian laws are not effective or sufficient. In order to curb these violations, there must be a specific mechanism to investigate them. Considering the Kurdistan region, we suggest that an institute is to be established to conduct research on landmines and to analyze collected and statistical data. Such an institute may have a major role in coordinating research related to different aspects of landmines such as information, prevention, training, disability, rehabilitation and the labour market. In addition, it makes it possible to gather evidences on the crimes committed to support the regional government in its efforts to follow up the landmine issue for matters of prevention, reduction of impacts, victim compensation and bringing the landmine producers, distributors and users to justice. We hope this book helps to bring positive changes to the landmine situation and conditions of the Kurdish people.

We acknowledge our indebtedness to friends who work in the mine action program: Essam Mhiaddin from iMMAP who prepared the images for the cover of this book and also provided many GIS data and maps to support the content of this book, Petri Nivalainen from iMMAP, Daniel Ericson from GICHD, and Dashti Mohammed from MAG for their support in providing the updated data, and for similar favours. Comments and suggestions from participants at seminars held at the University of Kurdistan Hawler, Soran University and Binghamton University are gratefully acknowledged.

Seoul July 2012

Almas Heshmati (Korea University and Soran University)
and Nabaz T. Khayyat (Seoul National University).

BIOGRAPHIES OF THE AUTHORS

Almas Heshmati

Almas Heshmati is Professor of Applied Economics at the Korea University and Research Associate at Soran University. He held similar positions at Seoul National University, University of Kurdistan Hawler, RATIO Institute (Stockholm) and MTT Agrifood Research (Helsinki). He was a Research Fellow at the World Institute for Development Economics Research, United Nations University, and Associate Professor at Stockholm School of Economics. He has a Ph.D. degree from the University of Gothenburg (1994). His research interests include applied microeconomics, econometrics, globalization, development strategy and performance analysis with application to manufacturing and services. In addition to more than one hundred scientific journal articles, he has published books on the EU Lisbon process, global income inequality, East Asian manufacturing, Chinese economy, technology transfer, information technology, sustainable development and power generation. He has also authored two books on water resources and the economy of Southern Kurdistan.

Almas Heshmati, Professor of Applied Economics, Department of Food and Resource Economics, College of Life Sciences and Biotechnology, Korea University, Anam-dong, Seongbuk-gu, East Building, Room #217, Seoul 136-713, Korea, e-mail: heshmati@korea.ac.kr, Phone: 0082-(0)10-4513 1712.

Biography of Nabaz T. Khayyat

Nabaz T. Khayyat was born in 1971 in Erbil, received a BSc in Mathematics (1993) from the University of Salahaddin Erbil, a High Diploma (1994) in Artificial Intelligence and Database Design from the University of Technology, Baghdad, and an MSc. in IT Engineering (2010) from the Seoul National University in South Korea. He worked for the UN for five years and different international organizations, and recently held a position with KRG as an Information Management Advisor in the Ministry of Agriculture and Water Resources. Currently he is studying for a Ph.D. in engineering at the Seoul National University in South Korea, and also studies for a Ph.D. in Economics at the Swiss Management University in Switzerland. His research is mainly on the adoption of mobile telephony, consumer service satisfaction, science and technology parks, technological capability, the effects of ICT on costs, water resources, Energy consumption, and landmine fatalities.

Nabaz T. Khayyat, PhD Candidate, College of Engineering, TEMEP, Seoul National University, San 56-1, Shilim-dong, Kwanak-gu, Seoul 151-742 Korea, e-mail: nabaz@snu.ac.kr, Phone: 0082(0)-10-7518-5575, 00964(0)-750-445-0486.

LIST OF TABLES

LIST OF FIGURES

LIST OF MAPS

CHAPTER ONE

INTRODUCTION: BACKGROUND AND HISTORY OF LANDMINES

Landmines were first used during the American Civil War in the nineteenth century and were referred to as “Land Torpedoes.” However, mines as they are known today were originally developed during World War I. Following the introduction of tanks to break the impasse of trench warfare, anti-tank mines were developed as an outcome of this new invention. Given the size of anti-tank mines, it was relatively easy for the enemy troops to enter minefields and remove the mines for their own use. This led to the emergence of the anti-personnel mine, a much smaller, delayed-action explosive device that was sown throughout anti-tank minefields to deter enemy soldiers from entering. It was first used to protect the more valuable anti-tank mines, but the anti-personnel mine has taken on a life of its own.

The landmine first became a prevalent weapon on the battlefield during World War II, during which it was estimated that around three hundred million anti-tank mines filled with trinitrotoluene (TNT) were deployed. One of the most effective anti-personnel landmines during this time was the German-made "bouncing betty" which was designed to jump from the ground to hip-height when activated and to propel hundreds of steel fragments within a wide range.

It should be noted that the main characteristic of a mine unlike other systems is its time explosive delay function. Mines are not been designed for immediate effect, and can stay inactive almost for an unlimited time, until triggered. This unique and inexpensive property therefore meant that mines were used intensively during the wars between nations and internal conflicts as offensive weapons to block access of the opposition forces to large piece of land.

During the time between the two World Wars, mines became easier to access and their design was modified. The Chinese and other nations manufactured mines at lower costs. Although they were made mainly from metal, plastic, ceramic and wooden landmines were also designed to be harder to detect.

1. Mines and Minefields

Mines can be defined as munitions to be placed under, on or near the ground or other surface area, and to be exploded by the presence, proximity or contact of a person or a vehicle. Landmines are usually very simple devices in terms of technology and readily manufactured anywhere. They are victim-activated and indiscriminate in their targets and impacts. Mines emplaced during a conflict against enemy forces may kill or injure civilians many decades later. There are two basic types of mines (King 1996): anti vehicle or anti-tank (AT) mines and anti-personnel (AP) mines. AT mines are munitions designed to explode from the presence, proximity, or contact of a vehicle as opposed to a person. They are comparatively large (consisting of 0.8 to 4.0 kg of explosive), often laid in unsealed roads or potholes, and detonate when a vehicle drives over them. They are typically activated by force (more than 100 kg weight), magnetic influence, or remote control. AP mines are munitions designed to explode from the presence, proximity, or contact of a person. AP mines are much smaller (80 to 250 g of explosive) and are usually activated by much lower force (only 3.0 to 7.0 kg weight) or tripwires.[1]

2. Impacts of Landmines

Worldwide landmine contamination problems can be considered a serious and hazardous issue for humankind. There is an estimation of one hundred million anti-personnel mines in the ground, and another one hundred million stockpiled around the world in spite of efforts made by various projects of the United Nations,[2] where more than forty-one million stockpiled anti-personnel mines have been destroyed, and their production, sale and transfer have been stopped.[3] Moreover, according to United Nations reports, if the use of mines were to be stopped immediately, it would take another one hundred years and the cost of more than $33 billion dollars to clear the existing landmines in use (ICRC 1997). The estimated cost measured in today's quantity in ground and price is actually much higher than those given above.

The existence of mines is considered a vital socio-economic and environmental problem facing many countries exposed to their use. Landmines and unexploded ordnance disposal (EOD) are considered significant threats and barriers to economic and social development in many parts of the world that face or engage in internal and external conflicts.

Prevention of access to the land for agriculture use, transportation, and water blockages have caused the displacement of people in many countries. Mines also cause the lack or low level of basic development infrastructure in those affected places. Beside the socio-economic impact, mines cause land and soil degradation, loss of biodiversity, and limits agriculture productivity (Berhe 2007). Solution to these problems and reduction of their impacts and consequences are the main purpose of the Mine Action activities.

Landmines pose a great risk of death and injury to civilians and wilds. The affected lands become virtually unusable for agriculture, transportation and socio-economic development. Landmines cause severe poverty and underdevelopment as landmines will, in general, block access to and limit most socio-economic activities in rural areas.

The effects of landmines on our lives are massive. They impede the ability of communities to recover fully from conflicts after the end of wars and conflicts. Beside the direct effects on life, landmines impose a heavy economic burden on these communities. For example, it costs an average of $600 to remove one mine, and around $1,500 to provide an artificial limb to survivors from mine accidents.

Other significant medical, psychosocial, political and economic impacts of landmines are many. For example, mines typically maim or kill the most productive members of a community's workforce, and prevent refugees and internally displaced persons from returning to their homes after the cessation of conflicts. As such, mines prevent a return of normal and productive life to rural areas and maintain economic, social and political pressures on urban life.

The effects of anti-personnel landmines disproportionately affect the poor and undeveloped countries. Poor rural inhabitants are often trying to grow crops on lands that are known or suspected to be landmine affected. They are forced to clear these lands from mines by themselves despite the lack of training skills and necessary equipment for that. They also lack from medical services such as first aid, prostheses if they become victims of those landmines.

3. Mine Action

The term "mine action" is used to describe:

> … all activities geared towards addressing the problems faced by populations as a result of landmine contamination. It is not so much about

> mines as it is about people and their interactions with a mine-infected environment. Its aim is not merely technical (to survey, mark and eradicate landmines) but humanitarian and developmental. The objective is to recreate an environment in which people can live safely, in which economic, social and health development can occur free from the constraints imposed by landmine contamination, and in which the victims' needs are addressed.[4]

The mine action process involves those functions and activities that focus on preventing the future use of landmines, reducing the risk of injury from existing mines and other ERW (Explosive Remnants of War) and assisting individuals' who have already been injured by mines and ERW. The prevention of the future use of landmines is carried out through advocacy to stigmatizing the use of landmines, spearheaded through the international campaign to ban landmines and through the process of mine stockpile destruction. Reducing the risk of injury for people in communities already contaminated with landmines and unexploded ordnance (UXO) is conducted through mine risk education (MRE) and demining programs which includes survey, mapping, marking, fencing and clearing. Victim assistance is the provision of assistance for survivors of landmine incidents, and includes rehabilitation and reintegration into their communities.

A significant part of this book will be devoted to Mine Action, a process concerned with "… reducing the social, economic and environmental impacts of mines and UXO with the intention of fostering subsequent human development" (Harrisonburg 2002). The action can be undertaken by national, international or non-governmental organizations. Severity of the situation determines the composition and cooperation among the organizations and the extent of their activities.

It should be noted that mine action is not only about demining operations. It is a comprehensive operation with multidimensional aspects with emphasis on people and societies and how they are affected by landmine contamination and its impacts. The objective of mine action is to reduce the risk from landmines to a level where people can live safely and in which economic, social and health development can occur free from the constraints imposed by landmine contamination, and in which the victims' needs can be addressed adequately.

3.1 Mine Action Activities

The mine action process comprises five complementary groups of activities: mine risk education; humanitarian demining or mine and UXO disposal surveys; mapping, marking and clearing activities and victim assistance which include rehabilitation and reintegration; stockpile destruction, and; advocacy against the use of anti-personnel mines. A number of other enabling activities are required to support the successful implementation of the process: assessment and planning; mobilization and prioritization of resources; information management; human skills development and management of training programs, and; quality management and the application of effective, appropriate and safe equipment used in mine action.[5]

Regardless of the context in which mine action takes place, it is rarely a standalone activity. Mine action is normally one of many activities required to take place in a post conflict environment. However, it is often the foundation for subsequent humanitarian and developmental initiatives. As such, it must also be an integral component of strategies designed to rehabilitate health care, education, infrastructure, agriculture and other systems required by societies in recovery from violent conflicts. Mine action' managers need to ensure that there is dialogue with different mine action participants from the wider humanitarian and development sectors as early as possible. This will ensure that the mine action program is properly focused on addressing the humanitarian and development challenges posed by mines and UXO.

Moreover, there are two types of demining activities: military and humanitarian. Military demining is the process undertaken by armies to clear a safe pathway to ease the soldiers' movement during conflicts, only clearing the land required for the troop's movement. Although the military lay the mines, humanitarian organizations with limited resources undertake the cleaning.

On the other hand, humanitarian demining aims to clear land for the sake of civilians in order to return to their lands and to perform their daily life activities without the threat of landmines and ERW. Once the land is cleared, a verification process should be undertaken in order to confirm the clearance of the land. Thus, the aim of humanitarian demining is to restore peace and security at the community level. In addition, humanitarian demining requires that the entire land area be free of mines. According to the UN mine clearance standard, a rate of 99.6% from the contaminated area should be cleared for humanitarian demining. It implies that for every one thousand mines removed from the ground, a maximum of four are missed which remain a threat to the civil population.

Humanitarian mine action operations face many complications on the ground. These include but are not limited to the following:

- Mines and UXO laid for many years may be decomposed, waterlogged, saturated with mud or dirt and can behave quite unpredictably. This will make it rather difficult to handle through normal demining processes.
- Stakes that carried fragmentation mines may have fallen over.
- Trip wires may be caught up in overgrown bushes, grass or roots.
- Wind gusts may sway bushes enough to pull a trip wire and detonate a nearby mine.
- Mines can be buried in deep ground to avoid discovery by metal detectors. Hitting a mine may simply dislodge dirt which allows mines to detonate the next time a person steps on it. Similarly, deeper mines may not detonate when the ground is hard, but rain may soften the ground to the point where even a child's footstep will set them off.

3.2 Humanitarian Demining Methods

The humanitarian mine clearance applies different methods to perform clearance on contaminated land area such as the following:

- Manual Demining: this method is using skilled and trained demining personnel to use metal detectors tools and thin prodders to find and detect mines in the ground. The detected mines are then collected and destroyed by controlled explosion.
- Mine Detection Dogs (MDD): this method uses trained dogs to detect the presence of mines and ERW in the ground by smelling the ordinance. MDD can detect mines and ERW with low metal and no metal content and mines and ERW in areas with high metal contamination or background such as on railway lines.
- Mechanical Clearance: this method applies specific machines such as flail, rollers, vegetation cutters and excavators attached to armoured bulldozers as the primary clearance method to remove and/or destroy mines and ERW from a given area to the quality of clearance laid down by the International Mine Action Standards (IMAS). These machines can only be used in certain terrains, and are expensive to operate. In most situations they are also not 100% reliable, and the work needs to be checked and reconfirmed by other techniques.

4. The Context in Which Mine Action Takes Place

The threat from landmines and other explosive remnants of war exists as a result of conflicts between or within countries. Landmines are first and foremost a humanitarian concern, and as such must be addressed from this perspective. They are also an impediment to the rehabilitation and sustainable development of societies. The nature of a conflict and the circumstances which exist in its aftermath will depend on many factors, and the form and extent of the required humanitarian and developmental assistance will vary accordingly.

There are considerable international debates on the forms, extents and phases of humanitarian and development activities in the post conflict activities, and there is considerable overlap between the various aspects of a country's recovery. These aspects include peace building, repatriation of refugees, rehabilitations of communities, reconstruction and development. For planning purposes, five general stages can be identified between the extremes of open conflict and stable self-dependency (see Fig. 1.1 below). Indeed, it is possible for more than one of these aspects to coexist within the same geographic area at a given time. Under such circumstances, local conditions may enable one part of a country to progress rapidly to assisted development or stable self-dependency, while open conflict may continue elsewhere in preventing the progress.

During open conflicts, landmines and other explosive remnants of war may significantly constrain the freedom of movement of any peacekeeping forces and/or humanitarian personnel and restrict their activities. MRE can be conducted to raise awareness of the danger and to promote safe behaviour in affected areas which is particularly important for refugees and internally displaced persons (IDPs) on the move. Small-scale mine clearance activity may also be conducted where there is an immediate threat of injury or loss of life during peacekeeping operations or in support of humanitarian aid work. However, this will be localized and of expedient character. The provision of assistance to those injured by mines will probably be needed. It may be possible at such an early stage to conduct data collection in areas that are known to contain mines and UXO. The data collection and gained information will play a crucial role concerning preparation for planning and subsequent more wide ranging mine action activities when conditions are permitted.

Fig. 1.1. General stages of post conflict humanitarian and development activities

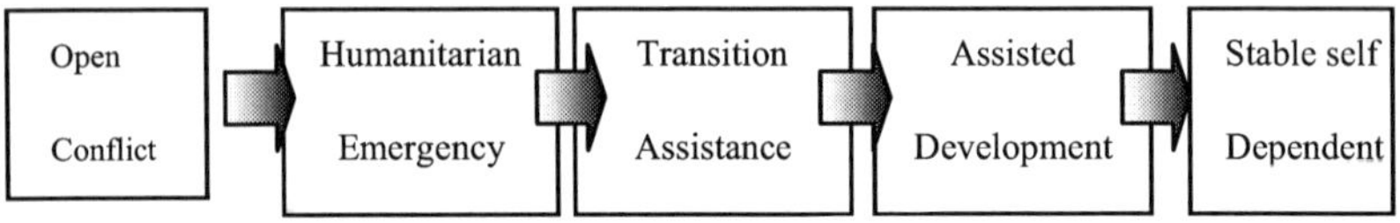

It is quite common that in the immediate aftermath of a conflict, often categorized as the humanitarian emergency phase, an effective governing authority may be absent and the conditions for stable and long lasting peace are therefore unlikely to be presented. The collapse of public and private organizations and subsequent ongoing civil unrest, a destroyed infrastructure, large numbers of displaced persons, deprivation and human suffering will attract international interests and the call for an urgent humanitarian response. Initially, mine action will focus on supporting emergency humanitarian and peacekeeping activities, rather than building indigenous capacity, although it may be helpful that in planning the former, the longer-term needs are to be taken into consideration. Although there may be much confusion and uncertainty at this early stage, it is vital that all forms of mine action activities are coordinated at the earliest opportunity. The scale and impact of the mine threat will become more apparent at this stage with an increasing demand for appropriate, effective and timely victim assistance together with a growing needs for mine risk education (in particular for displaced persons as they return to their homes in greater numbers). Therefore, survey and mapping of hazardous areas will be conducted and some areas may need to be marked and fenced.

A crucial period of transition exists between the humanitarian emergency and assisted development phases, during which the emphasis will shift from urgent humanitarian assistance to longer-term reconstruction and development activities. Projects may include the building of basic accommodation to shelter people, medical care and schooling and the rehabilitation of existing infrastructure. External assistance will still be needed for urgent reconstruction activities. However, during the transition phase emphasis will be placed on building an indigenous capacity to manage different needs that might be location specific. It is expected that the end of this period will be marked by the emergence of a recognized governing body demonstrating effective and democratic governance, and structures will be in place to provide sustainable funding for longer-term development programs. The mine action normally focuses on the development of an indigenous mine action authority, the collation and transfer of information and its management from various Non-Governmental Organizations (NGOs) and UN peacekeeping forces and the former warring parties, the establishment of enhanced data collection and needs assessment for mine action, the creation of a strategic plan and the training and deployment of MRE, survey, marking and clearance teams. The UN responsibilities at this stage will be to focus on the mobilization and prioritization of resources, the provision of an effective interface between the host nation, donors and mine action organizations, and the continued provision of guidance on standards, technology and quality.

A key characteristic of the fourth stage assisted development is the existence of a central governing authority with a clear mandate from the population, and with the growing confidence of the international community. By this stage, essential services and utilities are likely to have been restored, the fear of violence will have abated, opportunities for business and employment will start to improve, and the foundations for reconstruction and development will be in place for recovery to a state of stable local self-dependency.

International assistance under such circumstances will focus predominantly on the longer-term reconstruction and redevelopment programs, in particular those aimed at restoring and upgrading the country's destroyed infrastructure, services, agriculture and industry. By this stage, it is likely that the responsibility for mine action will be vested in the host nation, although depending on human resources some assistance may still be required from the international community. The priorities for mine action will be determined nationally but with advice from the UN and other key donors. Ideally, the mine action program will use local managers and local manpower, although it may be necessary to also use external agencies for the provision of specialist equipment, services and, if requested, quality control and financial audit.

Ultimately, a mine-affected country should be able to enjoy a stable and continuous stage from the development of conductive conditions of security, democracy and self-dependency without being hindered by the legacy of past conflicts. However, the extent of the problem may envisage that some assistance for mine action programs may still be required from the international community, although this will reduce gradually over time as sufficient local capacity is build up.

Clearly the form, characteristics and duration of each of these five general states will vary from case to case or conflict to conflict. Some countries may have great difficulty achieving the ultimate condition of stable self-dependency. Indeed, it is possible for a country to regress to a previous condition if the security situation deteriorates. If central government loses its authority or if international assistance is unable to be mobilized and sustained, a regressive development will emerge, leading to the escalation of conflict and prevention of continuous mine action activities.

This five general-stage model provides a perspective for analyzing the varying contexts in which mine action takes place. Humanitarian, rehabilitation and development assistance activities do occur simultaneously but the proportion of each activity will differ as a country progresses from a state of open conflict to stable self-dependency. This model may be useful as a starting point for the subsequent analysis of the conditions and for an academic work on the issue. It does not represent a template and should be applied intelligently, with care and with appropriate caveats.

5. Cooperation and Coordination

There are various stakeholders in mine action, including those internal to it (such as the UN and other international organizations, commercial and not-for profit implementing organizations, national mine action authorities, etc.), and external (such as donors, host governments, humanitarian aid and development organizations). The power and interest of each of the stakeholders will vary as the country progresses to the achievement of stable self-dependency, and as the mine action context changes in character.

It should be noted that regardless of the context in which mine action operations take place, it is rarely a stand-alone activity. Mine action is normally only one of many activities required in a post-conflict environment, but is often the foundation or a precondition for subsequent humanitarian, rehabilitation and development initiatives. As such, it must also be considered as an integral component of strategies designed to rehabilitate healthcare, education, infrastructure, agriculture, and other systems required by societies in the stage of recovering from violent conflicts. Mine action managers through their dialogue with non-mine action actors ensure that the mine action program is properly focused on addressing the humanitarian, rehabilitation and development challenges posed by mines and UXO.

Moreover, the mine action environment is quite complex. Mine action takes place in a variety of contexts and encompasses a range of activities, bringing together many stakeholders from different disciplines to cooperate to achieve the same stated goals. Its success depends on the ability of managers to coordinate and effectively plan the different activities. Management training is aimed at providing students with the knowledge and skills required to make the best use of scarce resources and work towards creating a situation where the social and economic impacts of mines and UXO is negligible, or at least up to an acceptable level of the affected communities.

The Kurdistan region of Iraq was massively contaminated by landmines, and this will continue to be a major obstacle for development and economic growth in rural border communities. The landmines were used systematically by the central Baathist government in its efforts in the sixties, seventies and eighties to depopulate the Kurds from the region and to weaken their demands for regional self-determination. The objective of this work is to investigate the magnitude of the mine problem in the region, the current mine action activities as well as the future clearance to facilitate rehabilitation and development programs. In addition, we emphasize the social, health, education and economic impacts of mine action and its legal consequences.

In order to achieve the abovementioned stated objectives, this work is divided into fifteen chapters. Following this introduction and background to the mine problem, we discuss in chapter two its international scope with some general statistics as background information. In chapter three we present the Kurdistan region's scope of the mine problem. We then review the methods of dealing with mines and their consequences in chapter four. The issues of mine action and impact survey, and factors influencing on the landmine impact surveys and the mine actions activities are discussed in chapter five to seven respectively. In chapter eight we review the outcome of the mine problem in the Kurdistan region obtained from the mine action and mine survey activities. We also discuss in more detail the nature of the problem and its characteristics by looking at some case studies in the region in chapter nine. We move to the assessment of the economic, social and health impacts of landmines in the Kurdistan region in chapter ten. Chapter eleven further discusses the human, security and political impacts of landmines in southern Kurdistan. The remaining chapters twelve to fourteen are allocated to discuss the issues of mine action resources and its progress, education and preventive measures, as well as legal actions against landmine users to receive compensations for clearance, life and properties and rehabilitation and development activities.

Notes

[1] See http://lm.icbl.org, Accessed on Aug. 2, 2009.
[2] See https://www.llnl.gov/str/Azevedo.html, Accessed on Aug. 4, 2009.
[3] For more details please refer to "Assistance in Mine Clearance," report of the UN Secretary General, Document A/64/267, August 12, 2009.
[4] Mine Action and Effective Coordination: The United Nations Policy. Resolution A/53/26 on Assistance in Mine Action adopted November 17, 1998.
[5] See http://www.mineactionstandards.org/imas.htm, Accessed on Aug. 12, 2009.

CHAPTER TWO

THE INTERNATIONAL SCOPE OF MINE ACTION

In this chapter we discuss the international scope of the mine problem and actions undertaken to prevent their use or to reduce their impacts primarily on innocent people. Despite growing international interest among non-governmental and governmental organizations to ban the use of mines, there is no interest in banning or limiting their production, in particular when anti-personnel mines are considered. These issues are discussed in this chapter and some basic statistics are presented to show a glimpse of the magnitude of problem and its effects.

The rest of this chapter is organized as follows. In Section 1 we discuss the condition in countries with mine problems. Section 2 provides quantitative information about the scope of the problem. Section 3 is allocated to the effects of landmines on humans. The issue of landmine ban treaty is discussed in Section 4. Finally, Section 5 gives a brief summary of this chapter.

1. Countries with Mine Problems

A numerous number of landmines were randomly and often laid by national elected/unelected governments or foreign countries with cross border conflicts to terrorize the innocent citizens of their local communities. Those landmines are mostly laid in the poor countries and in particular those of unstable political status and often against minorities in those countries. The exact number of landmines in use is unknown. However, it is estimated that millions of landmines and other UXOs are buried in countries and territories in all regions of the world. Thus, the devastation will continue unless political will and long-term funding change the situation (International Save the Children Alliance 2000).

A large number (around 120) of countries are reported to be contaminated by landmines or other explosive remnants of war ERW.[1] Such post conflict fragments create long lasting humanitarian problems with enormous social, economic and environmental consequences for the population of those contaminated regions.

The international organizations and communities such as the United Nations and the European Union are addressing tasks related to landmines problems in the world. These include mine clearance activities, reduction of the impacts of landmines on civilian daily life, helping the landmine victims and mine awareness campaign to prevent further placement of landmincs in areas inhabited by civilians and to reduce the mine incidents and its impacts. However, these countries and organizations do not pay attention to trade and the use of landmines by governments and international corporations. This issue will be discussed in more details in the coming chapters of this book.

According to the United Nation Children's Fund (UNICEF), there is one mine for every sixteen children in the world—in other words, one mine for forty-eight persons in the total world population, while according to the landmines victim's statistics, it shows that 800 people are killed and 1,200 are injured every month. This suggests that the magnitude and the impact of mines are larger in proportion to the number of mines in use.

Moreover, it is estimated that 500,000 landmine and explosive remnants of war survivors exist today, of whom three quarters are civilians. Thirty four percent of civilian casualties are children, nearly all of them boys. In some severely affected areas, children were among the majority of casualties: 59% in Afghanistan, 53% in Nepal and 66% in Somalia. Boys between five and fourteen are among the high risk groups who are subjected to the effects of mines (UNDP 2009).

2. The Scope of the Problem

Mines were laid on a wide scale in many regions around the world, with around four hundred million lain in different regions since the beginning of the World War II (UNIDR 1995). Furthermore, there is an increasing trend in their use in some countries and regions of the world.

Seventy-six countries are heavily affected by mines. Africa is the continent most heavily affected by landmines, with thirty-seven million lain in twenty-two countries, alongside fourteen countries from Asia Pacific, and other countries from the Middle East and North America (see Table 2.1 and Map 2.1). There are altogether no less than twenty-eight million landmines in Afghanistan, Angola and Cambodia standing for a large share of the total mine casualties.

Mines are used mainly because of their low economic cost and the ease of their use. However, the methods used to remove landmines are still slow, inefficient, risky and expensive. This has to do with lack of producing and using countries' interest in development of technologies to locate and neutralize mines. Therefore, the mine detection technology has not kept pace with rapid developments in mine production which have made them more deadly and more difficult to trace. Equipment designed in the World Wars is still being used to detect landmines produced in the 1990s. Most of the mines found in contaminated, mostly underdeveloped, countries were not made there, and such countries lack research into demining technologies.

A few countries like China, the former Soviet Union, and Italy were major producers and traders of landmines. Other important suppliers included the former Czechoslovakia and the former Yugoslavia, along with Egypt, Pakistan and South Africa. Prior to the mid-1980s, the United Kingdom, Belgium and the United States ranked among the top producers and exporters, other significant exporters in that period included Bulgaria, France and Hungary (Askin and Goose 1994).

3. The Effects of Landmines on Humans

The extensive use of Anti-personnel (AP) mines in recent years has caused a global humanitarian crisis. However, by counting the number of laid AP mines cannot accurately measure the problems caused by landmines. The effects can be measured in a meaningful way by estimating the high priority land where mines are hidden. Those lands could otherwise be used for agriculture, or pathways to access the basic facilities of human needs such as access to water from wells and riverbanks. These lands are limited for use due to risk of injury and death, and the land is therefore unused regardless of the amount of mines laid. Therefore, any attempt to count the number of mines around the world can be, at best, only an estimate. Thus, mine action groups focus on the humanitarian crises that landmines cause rather than only the quantity of them.

The deadly seriousness of the landmine story is in the numbers of people affected, especially the estimated number of new victims each year, exceeding thousands. Landmines cause huge restrictive barriers to social and economic development in underdeveloped regions of some of the world's poorest countries.

Since 1975, an estimation of more than one million landmines casualties have been reported, most of them civilians, and among them many children. Landmines do not always kill immediately, but can rather severely maim their victims, causing trauma, lifelong pain and often social exclusion. Worldwide, some 300,000 to 400,000 landmine survivors face terrible physical, psychological and socio economic difficulties.[2]

During the 1990s, there was an increasing awareness and activism around the controversial issue of landmines. Horrific stories and pictures from all around the world often showed that civilians were the main casualties in large numbers and this was set to continue in the future. Thus, there is a significant amount of inequality in the landmine threat and impacts.

According to the Landmines Monitor Report 2007, 5,426 mine victims have been reported worldwide. Despite of the presence of specific numbers of new mine victims, an accurate picture of the problem and its magnitude are not given. It can be inferred that landmines continue to cause losses in every region of the world. In 2007, in addition to the existing country cases, new infestations in thirty-seven countries and regions were reported in Africa, Asia, Europe, Latin America, and the Middle East. In addition to the hundreds of thousands of landmine survivors who were injured in the 1980s and 1990s, thousands of new victims each year suffer from injuries and permanent disabilities caused by AP mines. Moreover, livestock and other animals also fall victim, which can also be a major economic loss and problem.

The social cost of landmines is excessive but often under-estimated. Landmines can be found in the roads, pastures and crop farms, near drinking wells and river banks, and in residential houses and public places. As a result, they cause economic paralysis through restricting movement especially in the agriculture-based economies by reducing mobility of the productive and rural population.

Landmines affect human life in every aspect, including the ability of refugees to return to their homelands, cutting this fraction of the population off from the main source of income and living needs. Relocation of the population causes many social and economic problems in the establishment of shanty towns, unemployment and insufficient public services in urban area.

A landmine incident may cause various injuries to an individual including loss of limbs, abdominal, chest and spinal injuries, blindness, deafness, severe burns, and less visible, psychological trauma not only to the person injured in the incident, but to all the family members of those killed or injured.

Children are particularly vulnerable to landmines in a number of ways. If they are too young to read or are illiterate, meaning that signs posted to warn them from the presence of mines are of no use. Children are far more likely to die from their injuries compared to adults. Of those children who survive, few may receive prostheses that keep up with the continued growth of their stunted limbs. This discrepancy adds to their sufferings from landmine accidents. Many mine producers design landmines to look like toys to lure children rather than adults or military personnel. These toy-like mines were used intensively in Southern Lebanon in areas populated by people supportive to Hezbolla, the local Islamic political party.

Furthermore, the effects of landmines go further and even beyond what is observed and managed by the affected country's underdeveloped healthcare systems. The purpose of land mining activities is to depopulate certain targeted areas, to disrupt agriculture and to stop the flow of goods and services in those areas. The transportation system is also affected. The same affects will apply to agricultural and crop lands, religious sites, access to drinking water, national parks and forests and villages and the people living in those villages. It is obvious that all that which makes up the fundamental elements of a country can be affected by mines and will continue to destroy and kill innocent people. The producers' intensive research also makes them more effective in impact and less detectable. Mine impact surveys conducted by mine action activities provide the flow of valuable information to the producers to upgrade their technology and undertake measures to improve the performance of landmines in the form of higher impact and lower detectability. The producers may, for research purposes, finance mine action activities.

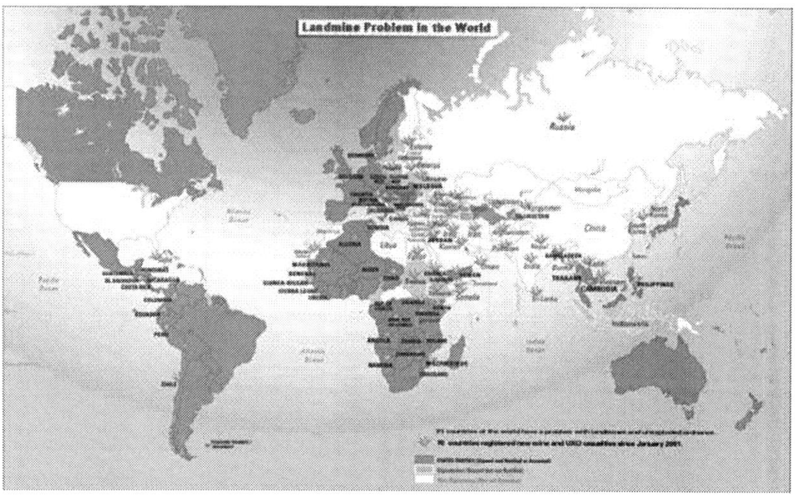

Map 2.1 Mine Contaminated Countries in the world

Table 2.1 below shows the number of states affected by mines in the different regions of the world. Africa tops the list with twenty-one countries followed by Asia (fourteen) and Middle East and North Africa (twelve). This table shows countries with mine contaminated areas that remained uncleared as of August 2009. Intensity in the degree of affection will be further-discussed later. Map 2.1 shows the mine-contaminated countries of the world.

Table 2.1. Mine affected states and other areas[3] (based on Landmine Monitor Report, 2009)

Africa	Americas	Asia pacific	Europe	Commonwealth of Independent States	Middle East and North Africa
Angola	Argentina	Afghanistan	Albania	Armenia	Algeria
Burundi	Chile	Cambodia	Bosnia and Herzegovina	Azerbaijan	Egypt
Chad	Colombia	China	Croatia	Georgia	Iran/***Kurdistan***
Democratic Republic of Congo, DRC	Cuba	India	Cyprus	Kyrgyzstan	Iraq/***Kurdistan***
DRC	Ecuador	Lao PDR	Denmark	Moldova	Israel
Djibouti	Nicaragua	Myanmar	Greece	Russia	Jordan
Eritrea	Peru	Nepal	Montenegro	Tajikistan	Lebanon
Ethiopia	Venezuela	North Korea	Serbia	Uzbekistan	Libya
Guinea-Bissau		South Korea	Turkey/ ***Kurdistan***	Abkhazia	Morocco
Mauritania		Pakistan	United Kingdom	Nagorno-Karabakh	Oman
Mali		Philippines	Kosovo		Syria/ ***Kurdistan***
Mozambique		Sri Lanka			Yemen
Namibia		Thailand			Palestine
Niger		Vietnam			Western Sahara
Rwanda		Taiwan			
Senegal					
Somalia					
Sudan					
Uganda					
Zambia					
Zimbabwe					
Somaliland					
21 states, 1 area	**8 States, 0 areas**	**14 States, 1 area**	**10 States, 1 area**	**8 States, 2 areas**	**12 States, 2 areas**

Table 2.2 below shows the mine victims in affected countries worldwide. The characteristics of countries and effects are expressed in the forms of population, number of casualties in 2008 and in the period before 2008 as well as the estimated mines per ERW survivors who received victim assistance. The table is in alphabetic order. In order to use it to compare the countries by their degrees of impact, one has to look at the ratio of victims per population or mines per capita population. Despite this limitation, the table is very informative and provides necessary information about the affected countries and furthermore allows comparison between the countries.

Table 2.2. Mine victims in the affected countries worldwide (Based on Landmine Monitor Report, 2009)

Mine Affected Country	Population[4]	No of Causalities before 2008	No of Causalities 2008	Estimated Mine/ERW survivors (received Victim Assistance)
Abkhazia	0.21	701	24	438
Afghanistan	12.07	811	992	52,000-60,000
Albania	3.17	18	0	510
Algeria	33.35	-	19	183
Angola	16.55	48	52	More than thousands
Argentina	39.13	-	-	-
Armenia	3.00	91	0	-
Azerbaijan	8.48	32	23	1,986
Bosnia and Herzegovina	3.92	30	39	3,919
Burundi	8.17	8	4	523
Cambodia	14.19	352	269	43,926
Chad	10.46	186	131	1,588
Chile	16.43	1	0	26
China	1,311.79	6,000	-	4,224
Colombia	45.55	895	777	> 6,163
Congo, Dem. Rep. of DRC	60.64	28	14	1,247
Congo, Republic of	3.68	0	25	-

Croatia	4.44	8	9	1,421
Cuba	11.26	-	-	-
Cyprus	0.77	0	6	7
Denmark	5.43	0	0	0
Djibouti	0.81	163	0	-
Ecuador	13.20	0	0	14
Egypt	74.16	25	62	645
Eritrea	4.69	70	64	84,000
Ethiopia	77.15	84	18	7,275
Georgia	4.43	-	35	591
Greece	11.14	393[5]	4	-
Guinea-Bissau	1.64	8	1	798
India	1109.81	170	33	1,903
Iran/***Kurdistan***	70.09	9,160	95	-
Iraq/***Kurdistan***	23.90	21,708	308[6]	-
Israel	7.04	2	2	13
Jordan	5.53	10	18	654
Kosovo	2.12	14	6	439
Kyrgyzstan	5.19	20	4	6
Lao PDR	5.75	100	100	7,000
Lebanon	4.05	130	28	2,720
Libya	6.03	13,852	10	3
Mali	11.96	11	14	7
Mauritania	3.04	618	14	>258
Moldova	3.83	66	3	-
Montenegro	0.60	277	3	260
Morocco	30.49	2,171	24	-
Mozambique	20.97	2,145	11	185
Myanmar	48.37	2,325	740	2,019
Nagorno-Karabakh	0.13	329	21	256
Namibia	20.04	618	11	473
Nepal	27.64	756	105	550
Nicaragua	5.53	1,236	3	1,145
Niger	13.73	80,035	89	157
North Korea	23.70	1	-	205
Oman	2.54	-	-	-
Pakistan	159.02	1,969	408	-
Palestine	3.83	794	37	654
Peru	27.58	318	8	372
Philippines	86.26	487	4	-
Russia	142.50	2,795	45	-
Rwanda	9.46	702	6	403–1,000
Senegal	12.07	702	24	570

Serbia	7.43	113	3	3,000
Somalia	8.44	2,354	116	1,405
Somaliland	3.50	814	59	489
South Korea	48.41	63	3	-
Sri Lanka	19.88	1378	84	1,158
Sudan	37.70	91	106	-
Syria/***Kurdistan***	21.22	600	11	330
Taiwan	22.92	100,0000	3	574
Tajikistan	6.63	802	19	448
Thailand	63.44	3,468	30	1,252
Turkey/***Kurdistan***	72.97	831	100	2,317
Uganda	29.89	900	20	1,000
United Kingdom	60.55	-	-	-
Uzbekistan	26.54	69	0	20
Venezuela	27.02	2	0	0
Vietnam	84.10	138,000	108	100,000
Western Sahara	3.77	2,500	50	450
Yemen	21.73	26	29	5,000
Zambia	11.69	62	11	152
Zimbabwe	13.22	1,550	7	1,300
TOTAL		1,307,096	5,501	

Like any other type of epidemics, landmine injuries should be treated. Epidemic treatments should be provided to the victims, with the prevention source of the epidemic found out. Here, the source is very likely the production of landmines. Therefore, a total eradication of this epidemic is possible through banning of the production, stockpiling, export and use of landmines. These measures are the most essential in the elimination of landmines, but very likely not convincing to the countries of production.

4. Mine Ban Treaty

The international Committee of Red Cross (ICRC), the non-governmental organizations (NGOs) in the International Campaign to Ban Landmines (ICBL), the UN and other countries have joined forces in a major campaign to focus the world's attention on the suffering caused by AP mines, mobilizing public support for a ban and stimulating a high level political and military support. This endeavour led to international negotiations to ban these weapons in 1996. A year later, on December 3, 1997, a number of countries and states signed the convention on the prohibition of AP mines in Ottawa, Canada. The treaty entered into force on March 1, 1999, following the fortieth ratification of the treaty.

The treaty imposes a total and unconditional ban on anti-personnel landmines was the result of negotiations led by a powerful and unusual coalition involving governments, the United Nations, international organizations such as the International Committee of the Red Cross, and over 1,400 NGOs through a network known as the ICBL. This unprecedented coalition used advocacy to raise public awareness of the impact of anti-personnel landmines on civilians and to rally global support for a total ban, the treaty can be summarized as follows:

- A total of 156 countries are states party to the Mine Ban Treaty.
- For the first time, there were no new ratifications or accessions during the reporting period.
- Some states who were not party to the treaty took steps consistent with its prohibitions.
- Non-State Armed Groups (NSAGs) have continued to pledge to observe a ban on anti-personnel mines. At least fifty-nine NSAGs have committed to halt use of anti- personnel mines over the past ten years. From 2008–2009, three NSAGs committed to halt the use of the weapon (the Moro Islamic Liberation Front in the Philippines, the Zomi Reunification Organization in India, and three factions of the Komala Party politically and militarily active in Kurdistan region of Iran).
- There are thirty-nine states which remain outside the Mine Ban Treaty, including some major stockpile, producers and users of anti-personnel mines, such as Myanmar, China, India, Pakistan, Russia and the United States.
- Three of the five permanent members of the UN Security Council (China, Russia, and the USA) remain outside the treaty members' list. The strong influence of these countries on decisions and their implementation reduce the success level of the activities.

Moreover, the international community, in adopting the mine ban treaty, conveyed the message to affected population that they have the right to freedom from the threat of landmines and the right to receive support in this respect. These rights have become preserved in international law. They created obligations for governments and organizations to provide assistance for mine action activities in affected areas, as well as for all actors in armed conflicts to cease the use of AP landmines. However, these rights and obligations will not be fulfilled without sustained political and financial commitments at the highest levels of government.

The Mine Ban Convention has also encouraged subsequent initiatives to protect civil populations from other explosive munitions that pose a serious threat. The adoption of two new international agreements as a result include the 2003 protocol on explosive remnants of war to the convention on certain conventional weapons and the 2008 convention on cluster munitions. Together, these three treaties provide a broad legal framework to protect civilians from weapons that kill innocent people long after the conflicts have ended.

5. Summary of International Scope of Mine Action

The landmine problem is not a phenomenon but an old issue which became more and more sophisticated as technology development allowed for further and profitable experimentation. Many countries of the world are also in conflict with neighbours and minorities within their territories with political, social and cultural demands that are not satisfied by undemocratic and often non-elected central governments. These governments use AP mines as effective and low cost measures in response to such legitimate demands and in their view as conflict resolutions.

In this chapter, we have discussed the international scope of the mine problem and actions undertaken to prevent their use or to reduce their impacts. The main objective was to pay attention to the conditions under which innocent populations suffer from their frequent, random and sometime calculated and planned uses. Despite growing international interest among non-governmental and governmental organizations to ban the use of mines, there is no interest in banning or limiting their production, in particular when AP mines are considered. In this chapter, these issues were discussed and some statistics presented to show the magnitude of the problem, its regional distribution and human and land use effects.

The combination of technology development and the high frequency of domestic and cross border conflicts have made landmines an international problem requiring global action to eliminate this inhuman phenomenon. We also discussed the strong movement against landmines, the progress of which is halted by major producers and users of landmines, namely the USA, Russia and China, all of which are members of the UN Security Council with veto, who do not hesitate to use it for their own national interests. This situation in many cases has proven to be a major obstacle in conflict resolution between countries and issues of a global nature. Inefficiency and the bureaucratic nature of existing international organizations dominated by the major powers adds to the handicapped situation in bringing necessary changes and in the achievement of noble goals to the well-being of humanity.

Notes

[1] Landmine Monitor Report 2008, Towards a Mine Free World.
[2] See http://www.mineaction.org, Accessed on Sep, 3, 2009.
[3] This table includes only states and other areas with confirmed mined areas. States with a residual mine problem are not included here, these countries include: Belarus, Honduras, Kuwait, Poland, Ukraine, and, Tunisia. The precise extent to which the Republic of the Congo, Mali, Moldova, Namibia, Niger, Oman, and the Philippines are mine-contaminated remained unclear as of August 2009. Both Argentina and the UK claim sovereignty over the Falkland Islands/Malvinas, Falkland Islands are mine-affected and therefore both Argentina and UK are included in the list. It is believed that both Djibouti and Montenegro have completed their mine clearance programs. However, this has not been formally confirmed so they remain on the existing list. Affected areas not internationally recognized as states are given in italics to distinguish them from other recognized states. Kurdistan is divided into Iran, Iraq, Syria and Turkey.
[4] The population of the countries under study is extracted from the World Bank Report in 2009.
[5] 291 illegal immigrants have been killed in Greek minefields.
[6] The figures are for the Northern Iraq mine action program (the Kurdistan region). No systematic casualty data collections are conducted in the southern and central parts of Iraq.

CHAPTER THREE

THE KURDISTAN REGIONS' SCOPE OF MINE PROBLEMS

Iraq in general and the Kurdistan region in particular are heavily contaminated with explosive remnants of war, landmines, and unexploded ordnance. Mines are related to the Iraq-Iran war, the Kuwait-related Gulf war, the invasion of Iraq by allied forces and the Iraq government's hostile assimilation policy. In the Kurdistan region, mines were laid during four distinct periods of its recent history. The first was in the mid-1970s, when the central government attempted to subdue the Kurdish resistance movements operating in rural areas. The second period was during the Iraq-Iran war, when border areas were heavily mined. The third period was during to the Gulf War, when barrier minefields were laid by the Iraqi government along the border with Turkey, and the fourth and final period was during the years since the end of the Gulf War, where the Kurdish factions in dispute have used landmines against each other.

Mines are mostly conventional type minefields. However, despite the fact that Iraq has recently signed the international anti-personnel mine treaty, so far none of the minefield records have been made available to the regional mine action program. It indicates an unchanged central government policy towards the Kurdish population. The mine action activity in Kurdistan was previously financed by the United Nations administered Oil for Food Program, and financed by some other aid programs. The Iraqi National Mine Action Authority was established in 2003 under the Ministry of Planning and Direct Cooperation, and later has been restructured under the Ministry of Environment. It is responsible for strategic planning and budgeting, project coordination, donor relations, national mine action standards, and maintaining the national Information Management System for Mine Action database. The transfer of financial, manpower and action responsibilities from regional authorities to central authorities has placed the Kurds in a vulnerable situation by making them dependent on decisions of a central government whose predecessors are in the first place responsible for the placement of mines. The objective of this chapter is to discuss these issues in more detail to shed lights on the scope of the mine problem in the Kurdistan region.

The rest of the chapter is organized as follows. In Section 1 and 2 we report some basic statistics on geography and population, and discuss the topography of the Kurdistan region. The scope of the mines problem in the region is discussed in Section 3, followed by a description of the organization, structure, responsibility, and the long run action plan of the National Mine Action Authority in Section 4 and 5. Section 6 is devoted to the conflicts in the Kurdistan region and the subsequent use of landmines against the civilian population. Section 7 is about the Mine Action activities in the region, and section 8 summarizes this chapter.

1. Geography and Population of the Kurdistan Region

The Kurdistan part of Iraq consists of five governorates, but currently only three are included in the Kurdistan region's territory. The three northern governorates of Iraq that constitute the region with a population of nearly five million are bordered by Syria, Turkey, and Iran, comprising a total area of 41,220 square kilometres, representing 9.5% of the total territory of Iraq and 17% of its population. The Region is geographically diverse, from warm plains into cooler mountainous areas blessed with natural springs where much snow often falls in the winter. The capital city of the Kurdistan region is Erbil, in which the Kurdistan Regional Government (KRG) is located,[1] locally known as Hawler. The citadel city in Erbil, with an approximately 5,000 year history, was prior to 2007 the longest continually inhabited city in the world. In recent years, it has been depopulated and renovation is in progress. The museum at the citadel is the most attractive tourist destination in the region.

As mentioned above, the Kurdistan region is located in the north-eastern portion of Iraq (see Maps 3.1A below), and in its current structure includes the governorates of Duhok, Erbil and Suleimaniya. It should be remembered that the Kurdistan region is not limited to these three governorates constituting regional self-governance. In addition, there is a significant Kurdish population in neighbouring governorates of Ninewah, Taamin, and Diyala (see Map 3.1B below), as well as the capital city of Baghdad. The Kurdistan Regional Government hopes to incorporate the disputed areas currently under the control of the neighbouring governorates.

Each of the three governorates is further divided into districts, sub districts, and villages. Within the Kurdistan Region, a total of twenty-five districts, eighty-eight sub-districts and 4,433 communities are officially recorded, of which 1,096 are mine-affected. The number of districts, size of population, the number of villages, and the land area of the three governorates are shown in Table 3.1 below.

According to the United Nations Development Program (UNDP) report (UNDP 2009), in the Kurdistan Region 3,512 minefields were officially registered. This covers an area of 788 square kilometres. To date, only one sixth of the mines lain have been cleared. According to the regional government figures, landmines and explosive remnants of war have claimed 8,174 victims (including both injuries and casualties) in the Kurdistan region between the years 1991 and 2007. In the coming chapters we elaborate with this issue in more detail.

Table 3.1. The Kurdistan Region's population size and area (2008 statistics)

Governorate	Total No of Districts	Total Population	Total No of Villages	Total Area km^2
Erbil	8	2,019,688	1,111	10,620
Dahk	7	973,118	917	9,920
Sulemaniya	12	1,917,936	2,405	20,680
Total	27	4,910,742	4,433	41,220

Source: Kurdistan Regional Statistical Office (www.mop-krg.org)

2. Topography

The Kurdistan region has four distinct geographic areas: The north-eastern mountainous region known as Al-Jazira rises to an altitude of nearly 2,135 m above the sea level, and is located near the Turkish border. The region's two highest mountains are Mount Halgurd, which rises to 3,728 m above the sea level, and Hasarost, at 3,600 m. The land area between the Tigris and the Euphrates rivers is an alluvial plain and is considered to be Iraq's most fertile region.

In the southeast, adjacent to the Persian Gulf, is a low-lying swamp area, containing marshes, lakes and reedy waterways. To the west of the Euphrates is the wide desert region, where the land gradually rises to join the Syrian Desert. This desert area constitutes about 35% of Iraq's total land area. The regional location of the Kurdistan region and its governorates is shown in Maps 3.1A and 3.1B below.

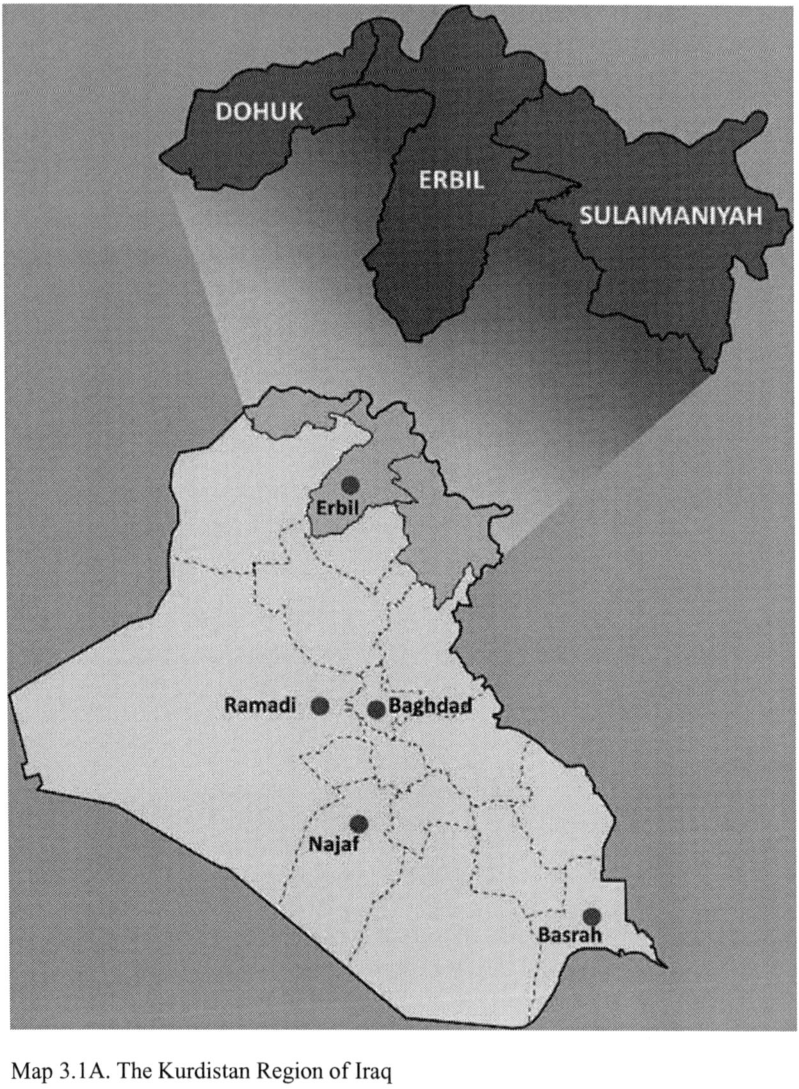

Map 3.1A. The Kurdistan Region of Iraq

Map 3.1B. The Kurdistan Region of Iraq

Iraq has two main rivers, the Tigris and the Euphrates. These rivers flow from the northwest to the southeast. They converge near the capital city Baghdad, then diverge, and meet once again about 160 km north of the Persian Gulf, to form the Shatt al-Arab river. This river flows through Basra and drains into the Persian Gulf. The rich alluvial soil characterizing the Tigris-Euphrates basin is an exception, as the soils here are very light and not especially fertile. The Kurdistan region with its fertile land and water resources has long been an agrarian society and a source of food production in Iraq.

3. The Scope of the Mines Problem

Iraq in general and the Kurdistan region in particular is heavily contaminated with explosive remnants of war (ERW), landmines and unexploded ordnances (UXO). Mines are related to the Iraq-Iran war (1979–1986), the Kuwait related Gulf war (1990–1991), the invasion of Iraq by allied forces (2003) and the Iraq governments hostility and assimilation policy (1960s onward until 1989).

The contamination affects the daily lives of individuals and communities, and impedes the delivery of humanitarian assistance, development and reconstruction projects at both the local and national levels. Prior to 2003, mine action efforts were concentrated in the Kurdistan region as part of the UN Oil for Food Program.[2] Following the conflict in 2003, these efforts were institutionalized under the National Mine Action Authority (NMAA) in Baghdad which was renamed the Directorate of Mine Action (DMA) in 2008. Regional Mine Action Centres (RMACs) were created in Erbil in the two mainly contaminated regions, namely the north and Basra in the south. In late 2006, the Government of Iraq signed the anti-personnel mine ban treaty, but the deteriorating security situation has remained a major challenge to carrying out necessary mine action activities and to fulfilling the treaty obligations.

Participants in the Iraq Mine Action Portfolio country team include the Ministry of Environment (MA), the UN Development Program (UNDP), the UN Children's Fund (UNICEF), a number of international and national organizations active in southern and central Iraq and the Mine Action Program in the Kurdistan Region including all national and international mine action organizations and UN mine coordination offices (UNMAS, 2009).

It should be noted that the countrywide threat to Iraqi lives and livelihoods varies from abandoned UXOs to dangerous surface contamination. Mines afflict vast rural sections of Iraqi land, with the largest areas stretching for hundreds of kilometres along Iraq's border with Iran. Large quantities of UXOs also remain scattered throughout cities and towns. These threats originated from minefields lain during the Iraq-Iran war in the 1980s, conflicts between rival ethnic and political parties, the military actions of 1990–1991 and the recent conflict of 2003 and its continuation. The foreign (i.e. American) forces are in a stage of leaving Iraq, but the political issues and disputes are still tense and not fully resolved. This might lead to the additional use of landmines in the future.

The Iraq Landmine Impact Survey LIS (VVAF 2008) illustrated the distribution of contaminated areas in thirteen governorates. The survey found 4,270 suspected hazardous areas and the contamination of more than 1,700 square kilometres of land affecting 2,117 communities. In the south, the two most common socio-economic blockages are irrigated land (87%) and fixed pasture land (91%). Rain-fed cropland blockages are rare, but migratory pasture blockages due to the tradition are more prevalent than in other parts of Iraq.

In the south of Iraq, two areas of threats are evident: (i) old mines along the Iran-Iraq border, and; (ii) new mines along the border with Saudi Arabia that account for about 50% of victims. UXOs cause almost as many injuries. Some areas have a deadly mix of ERW from the most recent conflict, including rockets and cluster bomb munitions. In the surveyed areas, 577 recent victims were documented in addition to thousands of older survivors. Altogether, it is estimated that more than 2.7 million people live in contaminated communities around Iraq.

It is vital to maintain the existing mine action resources in the Kurdistan region, and rapidly expand clearance activities and mine risk education programs in central and southern Iraq. Today, there are only a few humanitarian clearance organizations active in these regions, giving limited support to the DMA's target of a 50% reduction in ERW in five years. The insufficient level of activity in the south or reduction in their activities in the north more or less has to do with the internal organizational changes and disorganization in Iraq and the Kurdistan Region.

4. National Mine Action Authority

The National Mine Action Authority (NMAA) was established in 2003 under the Ministry of Planning and Direct Cooperation (MoPDC). Due to the transfer of responsibility of mine action from the MoPDC which officially took place in April 2008, the NMAA has been restructured under the Ministry of Environment as the Directorate of Mine Action. The DMA is responsible for strategic planning and budgeting, project coordination, donor relations, national mine action standards and maintaining the national Information Management System for Mine Action (IMSMA) database.

The DMA works with the Regional Mine Action Centre-South while the Iraqi Kurdistan Mine Action Centre and the General Directorate for Mine Action in Suleimaniya function independently in planning and executing mine action activities in the Kurdistan Region. A transfer of responsibilities from regional to central level might induce the lower efficiency and unfair allocation of mine clearance resources.

The UNDP is the leading agency of UN-supported mine action in Iraq. It periodically organizes the UN mine action coordination meetings to ensure that the UN's support is timely, coordinated and effective. The portfolio consultation and coordination process have been terminated at one of these meetings where all agencies were requested to share their information on NGOs working on mine action. Invitations for project proposals were sent to eligible appealing agencies (national authorities, national and international NGOs, international organizations, and UN entities with operational activities related to mine action and who appeal for funding through the portfolio). In 2003 a total of twenty-two projects from seventeen agencies were received.

The UN assistance strategy for Iraq 2008–2010 was signed in 2009 between the Government of Iraq and the United Nations. Under the strategy's protection sector, one concrete output for mine action was clearly spelled out: "Mine action policy, strategy and regulatory mechanism developed, Iraqi mine action operational capacity strengthened and awareness toward Mines and their impacts has been raised."

5. Long Run Action Plan

The mine action country team, which in the second half of 2009 has assisted the DMA in developing a draft for a national mine action strategy, has also supported the DMA in developing national mine action policies and management capacities to plan and execute sustainable mine action programs, with a view to addressing the development and socio-economic problems stemming from ERW.

Since all appealing agencies were in close contact with government agencies and/or the UN mine action team, the projects in the 2010 portfolio combined individual strengths under the framework of a comprehensive national mine action program. The government agencies include the Iraqi Ministry of Environment, the Iraqi Kurdistan Mine Action Authority (IKMAA), the General Directorate for Mine Action in Kurdistan (GDMA), and the Regional Mine Action Centre-South in Basra. The UN Mine Action Team involved several organizations including UNDP, UNICEF, the World Health Organization (WHO), the UN High Commissioner for Refugees (UNHCR), the UN Office for Project Services (UNOPS), the Office for the Coordination of Humanitarian Affairs (OCHA), and the UN Assistance Mission for Iraq (UNAMI).

The Directorate of Mine Action has indicated the following key objectives for its mine action program:

- Strengthening and developing the directorate's managerial capacities to coordinate mine action strategies and to meet the requirements for reconstruction and rehabilitation.
- Assessing the existing victim assistance capacities and developing additional capacities as needed.
- Establishing a nationwide victim surveillance system and mapping existing support mechanisms.
- Undertaking mine risk education in affected areas to raise awareness and reduce injuries and casualties.
- Reducing landmines and ERW impact through the development of national mine action capacities such as clearance and mine risk education provided by NGOs and commercial companies. To generate employment opportunities for the affected population.
- Reducing injuries and casualties by removing threats and raising awareness among the population.
- Encouraging the government to adopt national mine action legislation, international conventions and a legislative framework on landmines.

It should be added that that implementation of the abovementioned long-run plan will involve the following measures:

- Policy advice and institutional capacity building: This measure comprises providing advice and training for mine action institutions including the directorate and regional mines action centres, expanding national capacities to support humanitarian mine clearance and increasing the frequency with which priorities are determined and integrated with socio-economic components.
- Operational capacity building and coordination: Existing capacities need to be sustained at the nationwide level.
- Capacities in the centre and the south of Iraq require further development involving assistance to the DMA in identifying the needs of local NGOs, engaging international NGOs to help local groups to operate within the authority's strategic framework and initiating a strategy to transfer responsibility from NGOs to national institutions.
- Capacities for mine risk education: These need to be developed and sustained in the long run.

6. Conflicts in the Kurdistan Region

The troubled history of the Kurdistan Region has plagued the safety, development and prosperity of the Kurdish population. Mines were laid during four distinct periods of its recent history as follows:

- The first period was in the mid-1970s when the central government attempted to subdue Kurdish resistance movements operating in rural areas.
- The second period was during the Iraq-Iran war, when the border areas were heavily mined. These were mostly conventional type minefields. However, despite Iraq having recently signed the international anti-personnel mine treaty, so far none of the minefield records were made available to the mine action program. It indicates a discriminatory and an unchanged central government policy towards the Kurdish population's safety and equal treatment. The intensity of the conflict also resulted in a high concentration of battlefield ordnance or UXOs having been left behind. Minefields laid during the conflict that are within a five kilometre zone along the Iraq/Iran border are deemed "untouchable" by the government of Iraq, and the United Nations Office for Project Services (UNOPS) has been excluded from conducting clearance operations in this area since 1999.
- The third period was during the Gulf War, when barrier minefields were laid along the border with Turkey. Since the borders between Iran, Iraq and Turkey passes through Kurdistan, regardless of the legitimacy of aims like defence, it harms the Kurdish population.
- Finally, the fourth period was the years since the end of the Gulf War, where Kurdish factions have used mines against each other. These are mostly "nuisance" minefields, laid with no particular pattern, of which the engaged parties kept no records. More recently the two major political parties in the three northern governorates the Kurdistan Democratic Party (KDP) and Kurdistan Patriotic Union (PUK) have declared their commitment to abide by the Ottawa Convention and have signed the Geneva Call.

As in other conflicts, mines were used as offensive and psychological weapons to isolate or expel communities from their land and to deny access to farmlands, irrigation canals, power plants and roads. In many instances, once forced to flee from their homes, landmines have prevented the majority of disrupted people from returning to their homes and land to this day.

7. Mine Action in the Kurdistan Region

The populations of the Kurdistan Region face one of the most formidable landmine challenges in the world. Mines and UXO directly impact the lives of nearly 1,100 communities causing an estimated 160 casualties per year. Mine clearance, mine awareness and victim assistance activities within the region are conducted mainly by the United Nations Office for Project Services (UNOPS), nongovernmental organizations such as Mines Advisory Group (MAG), Norwegian People's Aid (NPA) or Solidarity. Later mine activities were conducted by the Regional Mine Action Centre (Iraqi Kurdistan Mine Action Authority, IKMAA), and augmented by the efforts of local villagers.

Furthermore, mine action was linked with other sectors of the United Nations humanitarian program in the Kurdistan Region. As of March 2003, it had paved the way for eight UN agency activities including housing, resettlement and the construction of electrification infrastructure throughout the region. It should be noted that the projects were financed by Oil and Gas revenues, part of which covered the UN organizations management of the different humanitarian and development programs.

The Oil for Food Program funded the clearance of some 76,500 landmines since 1998, restoring mine-free areas to landowners for agricultural and other civilian uses. Mine safety instruction courses were provided to 240 communities reaching 7,176 men, 8,353 women and 14,045 children, mainly in affected areas. Medical consultations for mine victims ranging from treatment to the fitting of orthopaedic prosthetics and rehabilitation are among other measures.[3]

The United Nations Office for Project Services UNOPS launched the ninth phase of its mine action project for the Kurdistan region in 2002. Its emergency demining program had a team of expatriate demining personnel stationed in each governorate where they trained local teams in the latest clearance techniques. As of August 2001, approximately 27,000 families in 165 communities have benefitted from UNOPS implemented mine clearance activities. As a result of this comprehensive work and program, more than four hundred additional tons of crops were produced in 2001 on cleared land, 34,700 livestock were able to graze safely, 3,300 people had improved water supplies and 400,000 people near the city of Suleimaniya gained access to electricity.[4]

The vision for the region, which UNOPS has developed with the many stakeholders concerned since the beginning of the mine action program 1997, was for a "Kurdistan Region free from the impact of landmines and unexploded ordnance by the year 2020." In order to meet the objective of this vision, for the period 2001–2006 the UNOPS' mission was to develop a local sustainable mine action capacity to provide effective relief support to communities affected by landmines, and to allow for the safe development of the local infrastructure.

UNOPS has managed the Mine Action Program (MAP) in the Kurdistan Region since its inception in 1997. The MAP was funded by the "Oil for Food" program as set out by the United Nations Security Council Resolution 986 and in the Memorandum of Understanding between the UN Office of the Iraq Program (OIP) and the Government of Iraq (GOI) in May 1996. The aim of the MAP was to address the impact of landmines and UXOs on communities by developing an adapted programmatic response, and by establishing the future direction of the program in order to support the sustainable development of the region. It should be noted that MAG and the Norwegian People's Aid did not come under the MAP but operated in parallel with it and as such it did not receive any financial support from the Oil for Food program.

The UNOPS Mine Action Program has adjusted its management and operational structure to meet the necessities of rapid program growth to support the development of sophisticated/adapted technology and to ensure the long-term sustainability of the endeavour. Its methodology centres on the needs of those communities most affected by the presence of landmines and UXOs.

8. Summary of the Kurdistan Region's Scope of Mine Problem

Iraq in general, and the Kurdistan region in particular, is heavily contaminated with landmines and unexploded ordnances, related to the frequent wars with neighbouring countries and the invasion of Iraq by the allied forces along with the Iraq government's policy towards the Kurdish population. In the Kurdistan Region, mines were laid during four distinct periods. The first was in the mid-1970s when the central government attempted to subdue the Kurdish resistance operating in the rural areas of the region. The second was during the Iraq-Iran war, when the border areas were mined. The third was during to the Gulf War, when barrier minefields were laid by the Iraqi government along the border with Turkey. The fourth was the years following the Gulf War, where Kurdish factions have used landmines against each other.

Iraq has recently signed the international anti-personnel mine treaty, yet so far none of the minefield records have been made available to the regional mine action program. The lack of minefield records has made the minefield clearance costly and time consuming, and indicates an unchanged central government policy towards the Kurdish population and its regional government. The mine action activity in Kurdistan was previously financed and managed by the United Nations administered Oil for Food Program and some independent Aid programs. However, The Iraqi National Mine Action Authority was established in 2003 and has taken the responsibility for strategic planning, budgeting, project coordination, relations with and between donors, setting national mine action standards and maintaining the national Information Management System for Mine Action database IMSMA. The transfer of financial, manpower and action responsibilities from regional authorities to central authorities has placed the Kurds in a vulnerable situation.

This reorganization has resulted in inefficiency, a lack of clearance activities and thereby the continued suffering of the civilian population in the region. The deteriorated situation is a direct result of the inability of the Kurdistan Regional Governments (KRG) to settle these issues in an optimal way in negotiation with the central government, and allocation of necessary resources financed from the regional budget and in parallel through formal legal claim efforts, such that the cost be covered by the central government budget.

Notes

[1] See http://www.krg.org, Accessed on Sep, 22, 2009.

[2] The Oil for Food Program was found in April 1995 as a temporary measure to ease the unintended consequences of the United Nations sanctions on Iraq's civilian population. The first Iraqi oil sold under the program aimed to pay for humanitarian supplies, was exported in December 1996 and the first shipments of food arrived in March 1997. As of 20 March 2003, the Oil for Food Program covered 24 sectors of needs and had prevented the further degradation of public services and infrastructure, making significant differences in the humanitarian situation nationwide. However, as expected, gradually it has generated a massive and bureaucratic organization with a large share of the population depending on it. It turned out not only bureaucratic but also corrupt and caused decline in local production as everything was imported.

[3] See http://www.undp.org, Accessed on Sep, 22, 2009.

[4] UN Office of the Iraq Oil for Food Program

CHAPTER FOUR

MINE ACTION COMPONENTS: A METHODOLOGICAL REVIEW

Mine action is the term used to refer to activities aimed at reducing the social, economic and environmental impacts of landmines and ERW. This chapter is devoted to discussing the humanitarian mine action activities. Humanitarian mine action in a general form is introduced in Section 1. In Section 2 the humanitarian mine action components are discussed, involving mine risk education, destruction of stockpile and victim assistance. A further disaggregation in Section 3 contains the components of mine risk education and mine risk avoidance. The mine risk educations component concerns objectives, activities and support for demining, victim assistance, advocacy and stockpile destruction. In Section 4 we discuss victim assistance, its definitions, objectives, scope of the problem, the reasons for mine accidents, and victim assistant components objectives, rehabilitations and its process. Finally, Section 5 covers the discussion of different landmine and ERW clearance surveys. It includes the general survey, technical survey, marking, clearance, and completion survey. Section 6 provides a brief summary of humanitarian mine action activities.

1. Humanitarian Mine Action

The problem of landmines and ERW is considered a serious issue against civilian people regardless of location and size. It is estimated that more than one million mine victims were either killed or injured since 1975, in which the majority were civilians (ICRC 2009). Landmines are designed to kill or cause serious injury through an explosive blast or fragmentation. A landmine victim suffers mostly from the shattering effect, and in addition it also causes an excessive medical and social burden. Furthermore, landmines, through contamination of useful and fertile agricultural land areas, also hinder the development of the affected communities and countries.

In addition to previous stocks of landmines, recent wars (including conflicts lasting only for few weeks) have left behind tens of thousands of ERW and UXOs worldwide, putting civilians at serious risk. Thus, mines and UXOs create long-term harm and destruction, in particular to rural areas with limited resources, and also influence the clearance activities and organization's priorities. The existence of mines and UXO in agricultural lands cause famines and the increasing need for food aid. Many people who live in mine and UXO contaminated areas are forced to choose either to stay and face the deadly continuous threat or move to other areas and hence cause production loss and added pressure due to the limited supply of housing and public services.

The impacts of landmines on individuals and communities go beyond physical damage. They harm the psychological, social, and economic status of individuals, families and their communities. By limiting movement and preventing children from the right to play, landmines erode peace of mind. During recent conflicts, landmines were intentionally used to target civilian populations, control their movements and damage their mental health to put pressure on the families and communities and impose their disruption. Moreover, the presence of mines and ERW continue to threaten and undermine peace, security and development in general. [1] According to the United Nations statement:

> Mine action is not just about landmines. In many countries, unexploded ordnance, or UXO, poses an even greater threat to people's safety. UXO comprises bombs, mortars, grenades, missiles or other devices that fail to detonate on impact but remain volatile and can kill if touched or moved. Some of the main sources of UXO are cluster bombs. Today, mine action programmes typically address problems of landmines, UXO and "explosive remnants of war (ERW)," which include UXO and "abandoned ordnance," or weapons left behind by armed forces when they leave an area.[2]

The increasing awareness of the global impact of landmines on civilians is a result of establishing humanitarian mine clearance programs. However, the programs need a well-coordinated global perspective. They have evolved at organizational, regional and national levels in different countries based on their individual requirements, while not much has been achieved internationally. Efforts are needed to organize the humanitarian mine clearance at international level and develop criterion, standards and procedures that generally and are internationally adoptable to enhance the effectiveness, cost efficiency, reliability and safety in the humanitarian mine clearance program in different parts of the world.

Moreover, landmines and other forms of unexploded ordnance create a lasting humanitarian problem with significant social, economic and environmental consequences. Mine action is multi-task and multi-objective in nature, involving those activities which prevent the future use of landmines (in terms of advocacy and stockpile reduction), reduce the risk of injury from existing landmines and ERW (mine risk education and demining) and in addition to that assists those individuals who are already injured (victim assistance).

The degree and form of landmine risk and ERW, and the nature of the mine action response, differ from one location to another in accordance to the specific character of the environment. The humanitarian mine action groups and organizations engaged understand such difficulties and adjust their priorities to the conditions. The availability of human and material resources determines the magnitude of their action response.

2. Humanitarian Mine Action Components

Mine action is the term used to refer to those activities aimed at reducing the social, economic and environmental impacts of landmines and ERW. It is not limited to de-mining (mine and UXO survey, mapping, marking and clearance), but covers also four other major components as follows:

- Mine Risk Education (MRE) also known as Mine Awareness (MA) which aims to help people understand the risks they are facing, identify mines and explosive remnants of war and to learn how to stay away from dangerous pathways.
- Victim Assistance (VA): Including rehabilitation and reintegration, medical assistance and rehabilitation services to victims such as job skills training (vocational training) and employment opportunities; in other words, providing them new skills so they can continue to be productive members in their communities. Assistance is also provided to families and communities who have lost members due to landmines.
- Destruction of stockpile: Help countries to destroy their stockpiles of mines as required by international agreements, such as the 1999 Anti-personnel (AP) mine ban treaty.
- Advocacy against the use of AP landmines: Advocating for a world free of landmines and encouraging countries to participate in international treaties and conventions designed to end and ban the

production, trade, shipment or use of mines and to endorse the rights of individuals with disabilities.

The four pillars mentioned above, along with the demining activities, construct the five interrelated components of the humanitarian mine action program.[3] All mine action pillars have the common goal to reduce the human, social, economic and environmental impacts of landmine contamination (See Fig. 4.1 showing the components of mine action and how they contribute to the reduction of risk).

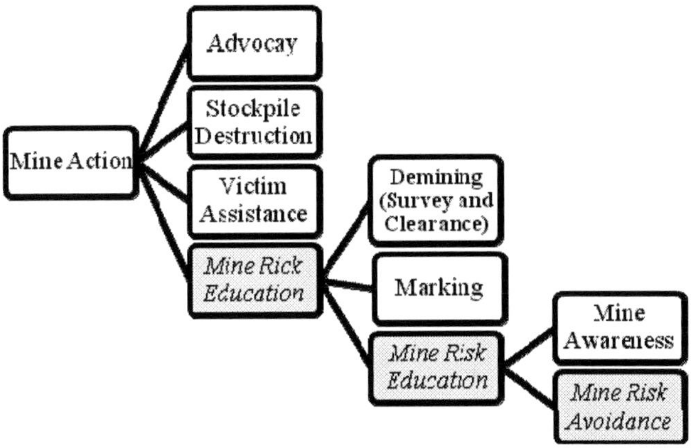

Fig. 4.1. Humanitarian Mine Action Components

3. Mine Risk Education

Mine Risk Education aims to reduce civilian casualties caused by mines and ERW through educating civilians to identify and avoid the risks associated with living in conjunction with a mine contaminated environment. Conducting mine awareness is particularly important when refugees and internally displaced persons return to their mine contaminated home towns pre-matured. The International Mine Action Standards (IMAS) defines the Mine Risk Education as the following:

> "Activities which seek to reduce the risk of injury from mines/unexploded ordnance by raising awareness and promoting behavioural change, including public information dissemination, education and training, and community mine action liaison."

It should be noted that MRE can be augmented to help in reducing mine-related mortality. It will contribute to ongoing mine clearance and raise the level of awareness among communities regarding mines, ERW information and training programs. It is worth to mention that the MRE activity goes beyond the concept of mine risk education; in addition, it seeks to prevent harm to civilians from all types of explosive devices. Thus, MRE covers the risk of ERW in addition to landmines.

The International Mine Action Standards IMAS has identified five different categories of risk takers (without being risk lowers in the context of business risk behaviour), and the reasons why certain individuals are disproportionally at a higher risk from landmines and ERW:

- The unaware: the person who knows nothing about the danger of mines and ERW: This category includes individuals such as children and refugees. It may also include illiterate people who cannot read instructions and information about danger in the form of signs. The majority of illiterate adult individuals are elderly and female.
- The non-informed; the person who knows that mines and ERW exist and are potentially dangerous but have no idea about safety procedures. This category of individuals includes those such people as internally displaced persons and older children (teenagers).
- The misinformed: the person who has received wrong messages and information. This category of individuals know about safety procedures, and are individuals such as ex-military soldiers.
- The reckless: the person who is aware of mines and ERW, and knows about safety procedures but ignores it intentionally. These individuals include adolescent boys who play with mines and ERW.
- The enforced: the person has little or no option but to intentionally adopt non-safety procedures. This category includes individuals such as adult in high mine and ERW affected communities who need to forage for food or water for their families to survive.

For a successful implementation plan, the MRE project should be integrated with other mine action component activities such as victim assistance, mine clearance and socio-economic development, and it should use the most appropriate mine awareness, community training and liaison methods designed and prepared for different target groups and target areas. Furthermore, it should target mainly the communities or areas most affected by mines and ERW where most accidents occur. It must also serve the basic needs of the most vulnerable groups in the society and those most at risk of suffering mine and ERW incidents in contaminated areas. This policy will reduce the frequency and magnitude of accidents as a result of higher risk taking implicated by satisfying the urgent need for the survival of families.

3.1 The Objectives of Mine Risk Education

The main objective of MRE is to minimize the mine and ERW risk for communities and individuals who live in mines and ERW contaminated areas. This can be achieved by developing their capabilities in using their own resources for MRE. In addition, the MRE aims to minimize deaths and injuries rates from landmines and other ERW. It also aims to reduce the socio-economic impacts of landmines and other ERW. In other words, the action supports the clearance process, victim assistance through physical and psychological rehabilitation and social reintegration of the survivors of mine and ERW. It should also undertake other activities such as the destruction of mine stockpiles and ERW by civilians in their homes and advocacy against AP mines, through supporting the AP mine ban convention and other international laws regulating landmines and ERW. Finally, it should support infrastructure development's activities in mine and ERW contamination areas.

3.2 Mine Risk Education activities

The MRE has three main activities that are usually introduced in chronological aspects but are in general undertaken in parallel. These activities are: Public information dissemination, Education and training programs, and Community mine action liaison. It should be noted that these activities refer to an external perpetuator and are conditional on support from a national government and its engaged organizations. In no means it includes cases of mine crime committed by the government on its citizens, such as the case of Kurdish population in Iraq.

Public information dissemination: Effective MRE involves communication among and between different individuals and different groups and communities, or "audiences." To achieve mine-safe behaviour it is important to not only inform and educate communities of safe behaviour but also to provide an environment that supports it. In cases with major mine problems, this could include legislations that support mine-safe behaviour or local/national political support.

In emergency post-conflict situations, due to time constraints and lack of accurate data, public information dissemination is often considered the most practical means of communicating safety information to reduce the risks facing the general population. Equally, this may form part of a more comprehensive risk reduction strategy within a mine action program, supporting community-based MRE and various demining or advocacy activities.

Education and Training: Education and training in MRE encompasses all educational and training activities that reduce the risk of injury from mines and ERW by raising awareness of the threat to individuals and communities and promoting behavioural change. Education and training is a two-way process which involves acquisition of knowledge, attitude and practice through teaching and learning. It is therefore more targeted to those people at risk. It uses more specific messages and strategies than is typically the case with public information dissemination. Moreover, education and training activities may be conducted in formal and non-formal environments through teacher to child education in schools, information shared at home from parents to children or from children to their parents, as well as from child to child education and peer to peer education in work and recreational environments.

Community Liaison: The main goal of community liaison is to place the needs and priorities of mines and ERW affected communities at the centre of the planning, implementation and monitoring of mine action and other sectors. Community liaison refers to information exchange between affected communities and national authorities, mine action and relief organizations, and development actors on the presence of mines and ERW and of their potential risks.[4]

3.3 Mine Risk Education support for demining

Through the community liaison activity, MRE can support the demining activity during its different operation stages (survey, marking and clearance). MRE teams and networks may gather and provide a range of important data and information essential for informing the planning and implementing of mine action activities.

MRE may provide the following information and support based on information supplied by the mine and ERW affected community:

- Information about the location of suspected/mines and ERW affected areas.
- Identify the types of ordinances presented.
- Undertake a complete assessment of socio-economic impacts of mines and ERW of the affected community.
- Information to support prioritization tools for marking or clearance.

- Supply of information on local warning signs in the affected areas.
- Encourage protection and respect for the signs and fencing already in the affected areas.
- Advise the community to support the demining teams.
- Inform the community of different safety procedures during the clearance operations.
- Inform community members of cleared areas and those that remain hazardous including proper markings of cleared and non-cleared mined areas.
- Support and participate in the land release procedures through building confidence for the community to make them trust that the cleared land is actually cleared and safe to utilize for agriculture.

It is important to post an impact MRE assessment by returning to the community within weeks or months after completion of clearance, and ensuring that the land is being used properly by the intended beneficiaries. Assessment of cleared land is required to ensure that the humanitarian objective given for the clearance procedure is reached and sustained. The assessment also allows for mine action organization to measure its humanitarian impact and effectiveness. This is vital in boosting moral and ensuring the continuation of donors' financial, technical and human resources support

3.4 Mine Risk Education support for Victim Assistance

MRE has a particular role in facilitating the provision of assistance to mine and ERW victims. For instance, a MRE team can do the following to support the victim assistance program during their work through visiting the affected communities:

- Identify national and local capacities for victim assistance.
- Identify individual amputees who need assistance and liaise with physical rehabilitation centres to ensure assistance is provided.
- Help in facilitating transportation of the amputees and family members from or to the medical centres for treatment.
- Consider employing survivors in their planned projects.

3.5 Mine Risk Education Support for Advocacy

MRE can play a significant role in advocacy against the use of AP mines through raising awareness in sessions, training, competitions and various public campaigns. It supports and advocates for victim assistance programs and enhances knowledge and skills through capacity building, including advocacy campaigns for relevant legal and suitable instruments as described below:

- Conduct awareness sessions in affected communities and districts.
- Coordinate activities between the district and national levels through monitoring and follow-up.
- Empower local persons through refresher training.
- Liaise and support the affected communities and victims.
- Conduct advocacy campaigns through seminars and workshops.
- Cooperate with other mine action organizations and participate in all relevant mine action forums.

MRE can also play an important role in building political desire in the concerned countries in favour of mine action activities. National and local ownership of the management of mine action is the only long-term, sustainable approach to dealing with the impact of mines and other ERW. This division of responsibility and coordination is dependent on the role played by the government in creating the problem. In optimal cases, this can be done through lobbying ministries and the parliament as well as generating public interest in, and support for, mine action through seminars and affective communication through the mass media.

3.6 Mine Risk Education Support for Stockpile Destruction

MRE activities may support the process of destruction of weapon caches and stockpiles and also explosive ordnance held by the government as well as those retained by civilians in their homes and farms. This can be considered both a process of information collection and advocacy of information collection to find out where weapons are stored or held, and advocacy to persuade families or local military forces to accept and ensure that they will be safely destroyed and not re-used.

4. Victim Assistance

4.1 Definition

Landmine Survivors are those individuals who are directly injured by a landmine explosion and have survived the accident. Landmine Victims generally refer to those who have been injured or killed by mine explosions and also their families who suffer emotional, social and financial loss as well as the communities that lose access to land and other resources. This more general definition is intended to recognize that the needs of those affected by the presence of landmines should determine the direction and magnitude of all mine action efforts.[5]

The convention's standing committee on victim assistance and socio-economic reintegration defines a "mine victim" as including the following categories of individuals and communities that incorporate all the items listed in a more detailed form above:

- Directly affected individuals.
- The families of directly affected individuals.
- Mine affected communities.

Additionally, victim assistance is a core component of mine action and an obligation of state parties under the Anti-personnel Mine Ban Treaty. Article Six, Paragraph 3 of the Treaty states that "Each State Party in a position to do so shall provide assistance for the care and rehabilitation, and social and economic reintegration, of mine victims and for mine awareness programs."[6]

Consequently, victim assistance is viewed as a wide range of activities that benefit individuals, families and communities. However, the term survivor assistance is sometimes used to describe activities aimed only at the individuals directly affected by a landmine incident.

In addition to what has been mentioned above, victim assistance includes, but is not necessarily limited to: minefield rescue, first aid, surgery, physical rehabilitation (physiotherapy and prosthetics for amputees), psychosocial rehabilitation and social reintegration of the survivors of explosions of landmines and other ERW activities.

4.2 The Scope of the Problem

The human and social impact of landmines and ERW continue to be a major source of humanitarian concern worldwide. Landmines disable and kill thousands of innocent people every year and can hold entire communities in a hostage situation. They prevent food and medical aid from reaching people in need, render agricultural land unusable, hinder the return of displaced families to their communities, and hamper already difficult relief operations and reconstruction efforts in many affected countries (UNMAS 2003).

The presence of mine problems many hinder the survivors from having access to some of their most basic needs such as food supply, access to water, adequate housing, roads, income, healthcare and access to the lifelong rehabilitation services that many community members urgently need. This means that mine survivors and other persons with disabilities are among the most impoverished groups in every mine contaminated society.

Hundreds of thousands of mine and ERW survivors exist in over seventy-five countries. According to the 2009 Landmine Monitor Report, there are up to 60,000 survivors in Afghanistan alone and over 45,000 in Cambodia. In 2008, the Landmine Monitor Report identified nearly 4,000 new injuries around the world caused by mines, explosive remnants of war and improvised explosive devices. While the actual figure is unknown, it may well be far greater, since many victims of mine accidents never reach a health centre and are therefore not registered.

4.3 Causes of Mine Accident

The most persistent causes of mine accidents can be attributed and identified by the followings:

- Lack of knowledge about mines/UXOs or the threat they impose.
- Economic necessity or survival imperatives: Many mine/UXO incidents occur because of deliberate decisions made by individuals or communities who may believe that the perceived benefits of preventive action outweigh the perceived costs of such action. For example, an individual might try to extract the aluminium component of UXO and sell it for much needed income. Another might wish to collect firewood in a mined area to use as energy for cooking or heating.
- Sudden environmental changes such as flooding may cause mines to become freshly exposed or shift their position.
- Belief in one's own invincibility: For example, people who have survived a long and bitter armed conflict may think that mines cannot harm them.
- Previously cleared areas have been re-mined.
- The recklessness engendered by familiarity among individuals exposed to mines or UXOs over long periods.
- Inaccurate or misleading information about mines or mined areas.
- Religious and cultural convictions favouring fatalism, or a belief in magic or in the power of karma. In some societies, people wear amulets in the expectation that they will protect them from injury in minefields.

- Behaviour consistent with one's self-image: For instance, men may be reluctant to report the presence of UXOs to mine clearance teams as this may be taken as a sign of their weakness or incapability.
- Curiosity can affect behaviour particularly among children. Learning about the dangers of mines/UXOs may even stimulate an individual's curiosity to find out more, such as: what will happen if the mine is touched? Or hit with a stone? Will it really explode?
- An individual may come under direct or indirect pressure from relatives, peers, friends, or other community members not to adopt safe behaviour. For example, peer group pressure may lead a boy to attempt to prove his courage by entering a minefield.
- Any combination of the reasons mentioned above can serve as the reason for mine accident.

4.4 Victim Assistance Components

Victim assistance generally comprises a number of individually or integrated activities involving the following components:[7]

- Pre-hospital care (first aid and management of injuries).
- Hospital care (medical care, surgery, pain management).
- Rehabilitation (physiotherapy, prosthetics appliances, and assistive devices, psychological support).
- Social and economic reintegration (associations and peer groups, skills and vocational training, income generating projects supports).
- Disability policy and practice (education and public awareness, disability laws, disability national level coordination body).
- Health and social welfare surveillance and research capacities (collection / processing / analysis / reporting).

4.5 Objectives of Victim Assistance

The determination of victim assistance monitoring is not connected with one country's comparison with another or to results at an international standard. The level of victim assistance is linked to the level of resources available in the country's custody. With regard to assistance, it is possible only to measure the financial, political and social effort of a country in relation to its socio-economic situation.

Thus, for each mine and ERW contaminated country, the following objectives are to be targeted with a given priority based on the resource availability conditions as follows:

- Evaluate the scale of the problem, in terms of time and space through recording and localizing the accidents and victims over a given time period.
- Evaluate the scale of the problem in terms of the difficulty of the injuries and the associated mortality rate.
- Measure the capacity of the affected country to deal with the needs of disabled people and landmine victims in terms of available facilities and services such as first aid posts, treatment and rehabilitation and prosthetic limb centres and social assistance as well as associations for disable people and mine victims.
- Evaluate access to health care for disable people and mine victims, with certain regard to social security systems and unpaid insurances.
- Examine the political and social environment, and legal measures available to disabled people and mine victims.
- Monitor the situation of disability and legal measures available to mine victims.

4.6 Rehabilitation of Landmine Victims

The majority of landmines and ERW victims are usually from mines and ERW affected countries where access to basic medical care is limited and the victims suffer from poor public medical care infrastructure, lack of systems for rehabilitation services, trained and skilled rehabilitation expertise, financial and technical resources and a vocational rehabilitation system. Other obstacles for rehabilitation include architectural barriers, lack of sanitation and other priorities for primary health care, high unemployment and a low societal awareness of disabled people's needs and capabilities.

According to the World Health Organization (WHO) report in 2003, most developing countries have some rehabilitation services. However, these services generally counted for less than 5% of the disabled people in the average community. These services are frequently based in general hospitals in centralized urban areas which results in prohibitive costs and limited accessibility for the general population with such needs, and are often inaccessible to those in rural areas.

4.7 Victim Assistance's Process

Assistance to landmine victims is a comprehensive process that includes medical, psychological, social and economic components. It aims at the full socio-economical re-integration of landmine victims into the society. As such, it is an integrated part of the overall system of assistance for persons with disabilities. The government has the responsibility of providing public assistance to persons with disabilities, including landmine victims, in the frame of their overall social and healthcare policies.

Within the victim assistance program, the landmine survivors will receive the same opportunities in general life, in healthcare, access to publicly provided social services, income earnings, education, and participation in community activities as anyone else in the society. The ultimate goal of survivor assistance programs should be the complete rehabilitation of mine and ERW survivors and their full reintegration into the wider community.

The process of victim assistance can be summarized as follows:

- Medical treatment: after the mine and ERW survivor is rescued and then received the first aid, he/she is likely will need medical treatment depending on the extent of the injuries sustained.
- Physical rehabilitation: this service is usually offered by rehabilitation centres (it may be attached to the mine action program of the country) which will determine the kind of services that are required. However, it's vital to insure that the assistance given can be sustained by the survivor, for example, wheelchairs should not be given to people living in hilly or rocky areas with few roads or paths; and prosthetic limbs should be simple, sustainable, affordable and available to the victims.
- Socio-economic re-integration: after the physical recovery, the victim assistance program will provide socio-economic assistance in the form of training, work cooperative and income generating schemes.
- Psychological care for the survivors and their families: the program will ensure provision of social support to the survivors and their families by giving training in different approaches such as counselling by using participatory approaches to highlight tradition methods, exploration of local attitudes to disability, understanding needs of disable patients, and finally methods of offering assistance. Moreover, the program can continue offering support through

assessment and provision of materials to locally based sources of traditional support.

5. Landmine and ERW Clearance

In this section, a detailed description will be given about landmine and ERW clearance steps taken and methodologies employed in the complicated clearance process. The process involves a general survey, a technical survey, marking, clearance and a completion survey. Each of these steps is described below in a more detailed form.

5.1 General Survey

The general survey (level one) is a process to identify, map and record areas contaminated by mines and ERW to quantify the type, size and scope of the mine problem to be dealt with. The general survey provides an overall understanding of the mine problem in a particular area, and assists in effective planning and efficient allocation of available and scarce mine action resources.

Moreover, a general survey is a deductive process intended to provide an approximate guideline to the extent of contamination by mines and ERW. When socio-economic factors are included, the impact of the contamination can be assessed. The core activity for the general survey is therefore a community-based data gathering process which is intended to gain basic information about the location and impact of landmines. This community data gathering process can be augmented by other data sources such as hospital records of casualties and historical information about the conduct of the conflict.

When the abovementioned information is processed using a geographic information system (GIS), it is possible to digitize the locations of reported contamination and victim incidents on a local map. This can then be compared with the information on population densities and development needs, based on which priorities for clearance can be established. The GIS system also acts as the basis for historical archiving allowing for future development to find out what type of clearance has already been done in a particular area.

The level one survey which is usually conducted by MRE teams is aimed at collecting information on many aspects of landmines in the affected community and the impact they may have on people's lives. In each community, data collection is conducted with representative groups of principally informants. Methods used may include community mapping and visual verification of suspected areas. The survey must be signed and certified by the leader in each community prior to its completion. The level one survey usually collects the following information:

- Information on each mine and ERW suspected area.
- Detailed information on victims from within the past two years as an indication of severity of risk.
- General information on victims from incidents of more than two years.
- Past and current mine action and basic community information.

Moreover, the survey data facilitates the program cycle of a feasibility study, needs assessment and subsequent planning components for necessary mine action program to be undertaken.

5.2 Technical Survey

The technical (level two) survey is a detailed technical and topographical investigation for known or suspected dangerous mine contaminated areas that have been previously identified during the first level general surveys. The aim of conducting a technical survey is to collect sufficient information for effective use of mine and ERW clearance resources by accurately identifying the areas to be cleared, marking (fencing), the depth of clearance, establishment of local soil conditions and the vegetation characteristics of the areas.

The information obtained from a technical survey will be used as the technical specification for the planning and management of a subsequent defined clearance tasks. The output of a technical survey includes perimeter marking to reduce the risk of unintentional entry into the suspected or identified dangerous area. This process is normally considered part of a comprehensive mine awareness and education program. If clearance does not immediately follow a technical survey, then survey markers should be left securely in place. Such markers will enable the accurate and safe location of the dangerous area at a later time.

Therefore, it can be inferred that the technical survey aims at detailed and precise marking, mapping and recording of minefields. It contributes to increasing the effectiveness of mine clearance activities through the preparation of areas for clearance operations. Minefield survey and planning plays a key role in the effective and efficient conducts of mine clearance operations.

The following information is collected during the process of the technical and topographical survey:

- Details of the suspect/hazardous area, typically the extent of the contamination, mine patterns, number of rows in which mines are planted, density of mines, and types of mines (AP/AT).
- Other hazards such as sub munitions.
- Boundaries of hazardous areas.
- Restrictions and limitations, such as proximity to dwellings, roads and different infrastructures.
- Details of the terrain, the vegetation and climate conditions.

5.3 Marking

The marking of mine and ERW areas is undertaken to provide clear and unambiguous warning of danger to the local population in order to reduce the risk of unintentional entry. Temporary marking systems may also be used to mark the perimeter of a dangerous mine and ERW area if it is confirmed the clearance will continue a short time after the completion of the technical survey. Marking mine contaminated areas is a safe procedure for productive use through the application of safe and efficient clearance and destruction technology to remove the threat of mines and other explosive devices. This is the component where most of the mine action resources are required that physically eradicate the threat of mines.

According to the International Mine Action Standards (IMAS) there are three general categories of marking systems:

- Permanent marking systems should be used to mark the perimeter of hazardous mine and UXO areas not scheduled for clearance in near future. They should employ a combination of markers and signs. In practice, only physical barriers such as fencing have proven an ineffective permanent marking measure. However, other less attractive physical barriers may be used if available in conjunction with markers and signs.
- Temporary marking systems may be used to mark the perimeter of a hazardous mine and UXO area in preparation for clearance operations. They may include the use of physical barriers of a temporary nature.
- Improvised marking systems are found to be generally more effective in places like Mozambique where a combination of mine

signs and painting trees with red paint have been used with relatively good outcomes.

5.4 Clearance

The aim of the survey and clearance operation is to remove the threat posed by mines and ordnance from suspected areas to facilitate their future use. The clearance operation will involve one or a combination of the following assets:

- Manual Clearance Teams (MCT).
- Mine Detection Dog Teams (MDD).
- Mechanical Assets such as:
  - Armoured dozers for creating access and safe lanes.
  - Armoured flail tractors for mine clearance, ground preparation, and vegetation removal.
  - Armoured excavators with flails/tillers, tines, and special sieve buckets.
  - Remote controlled wheeled and tracked vehicles for mine clearance and ground preparation.
  - Hand held mechanical blowers.

In addition to the MCT and MDD teams, internal quality assurance (QA) or quality control (QC) Teams also play vital rules. The purpose of QA in humanitarian demining is to confirm that management practices and operational procedures for demining are appropriate, and achieve the stated requirement in a safe, effective and efficient manner. Internal QA will be conducted by demining organizations themselves, but external inspections will be carried out by external monitoring body conducted, and the QC teamwork will be associated to the inspection of the "finished product"—safe and cleared land, highly useful to the community.

Within mine action, the quality of the process and the end product is assured through a two-stage approach as described below:

Stage 1: Quality assurance involves the accreditation and monitoring of the demining organization before and during the clearance process. To achieve this, demining organizations need to establish an effective management organization, develop and maintain different procedures, and in practice apply these procedures in a safe, effective and efficient manner.

Stage 2: Quality control involves the process of an inspection of cleared land before it is formally released to the beneficiaries for their use. This combined application of quality assurance (before and during the clearance process) with post clearance quality control contributes to achieve an acceptable level of confidence that the land is safe and free of mines for its intended use. The quality of clearance must be acceptable to both the national mine action authority and the local community, and needs to be measurable and verifiable.[8]

Landmine clearance is a complex process. Mines are usually laid in residential (urban and rural) areas, agricultural land, grazing land, irrigation canals, roads, etc. A single approach or one type of technology cannot be applied for the clearance of all types of terrains. Therefore, a "tool box" approach has to be adopted and different procedures and technology have to be used for the clearance of different types of land according to its terrain.

The current humanitarian mine clearance technology used in various parts of the world is based on the use of very basic equipment, techniques, and procedures that are developed by time. However, the use of this simple technology with proper procedures has proven to be practical, cost effective, and reliable. The Kurdistan region could be a good example of the use and evolution of this type of technology, which should continue and be further improved and expanded.

The current three main approaches to the humanitarian mine clearance are the following:

Manual clearance with the support of metal detector and prodder: In this method, mines are searched with the help of a metal detector and investigated with a prodder or bayonet to identify if the reading is only a metal fragment or a mine. If only a metal fragment, it is removed from the area to prevent further signals, and if proved to be a mine then it is subsequently destroyed by detonation in site with the help of an explosive charge. This method is useful for the clearance of areas mainly contaminated with anti-personnel landmines. Despite being a slow process and associated with safety concerns for the mine clearance personnel involved, this method has proved to be practical and very reliable.

Manual clearance with the support of explosive detection dogs (MDD): In this method, mines are detected with the help of explosive detection dogs. When a dog gives an indication, it is rechecked with a highly sensitive metal detector and then manually investigated. Fragments and mines are disposed of in accordance with the method explained above. This method is particularly appropriate and useful in the clearance of areas with a low probability of mines and areas contaminated with minimum metal mines. This method has proved to be relatively quick, but the safety concerns are still there as an important issue to consider.

Mechanical Clearance: Various types of high technology machines are used in some parts of the world to clear mines. However, they cost much more than the manual mine clearance and are not very reliable. In addition, the maintenance of such machines is extremely difficult in mine-affected countries. Therefore, these machines have been replaced in many contaminated countries by simple soil excavators (backhoes) with some modifications. They have proven to be very effective in the clearance of residential areas and irrigation canals where mines are buried up to two metres under rubble and soil. These excavators can further be improved and used on a large scale for the clearance of residential areas and irrigation canals.

All the above-mentioned methods of humanitarian mine clearance are effective and practical in different and specific types of terrain. Therefore, one single method cannot be used for the clearance of all types of terrain. An assessment of optimal methods is to be made prior to the actual clearance actions. The reliability, cost and its suitability are among characteristics that need to be considered by the mine clearance groups.

5.5 Completion Survey

Once the land is cleared, a completion survey (Completion Report) is conducted prior to the handover process. The survey collects information about the number of mines used, and number of ERW collected and destroyed mines during the clearance, an assessment or information whether the land is fully cleared or only partially cleared is conducted.[9]

In order to address the landmine problems from a humanitarian point of view all the interrelated aspects of mine action are to be taken into account. Even though the current technology used for humanitarian mine clearance is slow and not absolutely safe, it is still practical, reliable and affordable. In order to enhance productivity, cost efficiency and safety of humanitarian mine clearance, attention has to be paid to the development of affordable, easy to operate and field oriented technology. The efforts made in various parts of the world to address the landmine problem need to be coordinated, and the experience and expertise gained in this field has to be exchanged. Sufficient funds are to be made available for the ongoing humanitarian mine clearance operations so that the high priority areas are cleared as soon as possible to facilitate the return of refugees and internally displaced persons, allow for the resumption of necessary socio-economic activities and minimize death tolls caused by landmines.

6. Summary of Mine Action Components

Landmines and ERW are serious threats to civilian populations regardless of their locations. The number of landmine victims is large and growing due to the intensity of landmine use and technology developed in making them undetectable, as well as population increase and increased demand for food and energy. Landmines are designed to kill or seriously injure. Through contamination of useful and fertile land areas, it is also an obstacle to the development of a country.

Landmines and UXOs put civilians at risk of death or serious injury and create long-term harm to rural areas with limited resources. The existence of mines and UXOs in agricultural lands can cause famines, leading to an increased need for food aid. It affects the concentration of population in urban areas and subsequently creates a public service supply problem, resulting in the massive migration of the rural population.

The impact of landmines on individuals and communities is not only physical—it also affects the psychological, social and economic status of individuals and their communities. Children suffer most through the limitation of their movement. The increasing awareness of the global impact of landmines on civilians is a result of the establishment of humanitarian mine clearance programs. Efforts are needed to organize the humanitarian mine clearance at the international level and to develop criterion, standards and procedures that are internationally adoptable to enhance the effectiveness, cost efficiency, reliability and safety in the humanitarian mine clearance programs.

Landmines and other forms of unexploded ordnance create a lasting humanitarian problem with significant social, economic and environmental consequences. Here, the objective was to shed light on the mine action activities. Mine action involves those activities which prevent the future use of landmines (advocacy and stockpile reduction), reduce the risk of injury from existing landmines and ERW (mine risk education and demining) and assist those already injured (victim assistance). The degree and form of risk from landmines and ERW and the nature of the mine action response varies from one location to the other according to the specific characteristics of the environment.

Notes

[1] See http://www.womenwarpeace.org/node/9, Accessed on Dec. 23, 2009.
[2] See http://www.mineaction.org/section.asp?s=what_is_mine_action, Accessed on Aug. 22, 2009.
[3] Here, the word "mine clearance" has been progressively replaced by "mine action" which covers all the essential components of humanitarian mine action.
[4] IMAS Mine Risk Education, best practice guide book 6. 2005.
[5] See http://www.womenwarpeace.org/webfm_send/557, Accessed on Dec, 23, 2009.
[6] See Mine Ban Treaty, convention on the prohibition of the use, stockpiling, production and transfer of anti-personnel mines and their destruction, 1997. This document is available at:
http://www.un.org/Depts/mine/UNDocs/ban_trty.htm, Accessed on Aug, 22, 2009.
[7] Handicap International: Victim Assistance, Thematic Report 2000.
[8] For more details on quality issues, see The Swiss Foundation for Mine Action—Quality Assurance and Control.
[9] In some cases, the community is in urgent need of the use of existing mine-contaminated lands for some vital purposes such as farming or access to water for drinking. Therefore, landmines may sometimes be partially cleared and handed over to the community.

CHAPTER FIVE

THE LANDMINE IMPACT SURVEY

The tremendous, worldwide threat of mines and the impossibility of clearing all mine contaminated areas in all affected countries, these countries have to allocate efficient internal resources to prioritize the local activities of mine clearance according to local needs. The national priority activity can be achieved through conducting a general landmine impact survey (LIS) to obtain the necessary information about the magnitude of the problem and resource requirements for clearance. It also supports the decisions and involvement of various NGOs in such activities.

A landmine impact survey assesses the impact of explosive contamination on nearby communities. The main purpose of conducting LIS is to assess the landmine situation in each mine and ERW affected community, and to develop an understanding of how this threat affects the community and how its negative effects can be reduced or eliminated. LIS collects accurate, reasonable and useful data about mines and ERW. In addition to this preliminary precondition, LIS aims to support further steps in the mine action process including marking, clearance and completion surveys.

The rest of this chapter is organized as follows. In section 1 the purpose of a landmine impact survey is outlined. In Section 2, types of mine suspects including areas and communities are presented. Section 3 contains discussion of the global landmine survey. Section 4 discusses the survey process followed by a summary of the chapter in Section 5.

1. Purpose of Conducting a Landmine Impact Survey

A Landmine Impact Survey is a process that collects expert opinions on the presence and severity of a landmine problem and evaluates it through three basic rules for best understanding the problem faced and options available for mine affected communities:[1]

- All suspected communities are to be visited for assessment.

- During visits, key informants, i.e. people who live in those communities and have better information about landmines in their area, are asked about the preliminary identification of mine affected communities nearby.
- From the communities deemed mine-free, a sample is drawn. These communities are also visited. If any are found to be affected, other supposedly mines-free communities in their vicinity are also visited.

For adequate coverage, good comparability among country surveys and limited scarce resources, sampling plans should assure with 90% confidence that not more than 10% of the non-suspected communities in any affected district are in fact infested. Note that the 10% probability applies only to poorly known districts, not to a national average including mine infected and mine-free regions.

If that requirement leads to an impractical sample size, consultation with mine survey agencies is needed. The universe of sampling is the set of all communities deemed mine-free in a given affected district. It should be noted that, sampling is not done in districts on which there is complete consensus that they are mine-free. This is important for both statistical and community confidence building measures.

LIS strives to give as accurate a representation as possible of the number and distribution of mine-affected communities in the surveyed countries. This information is extremely useful in the subsequent resource allocation and priority formation. The methodology defines three processes that work together to achieve the desired accuracy. These include; (a) expert opinion is systematically elicited on communities that are likely affected, (b) all of these suspected communities are then visited, and finally (c) a quality assurance procedure adapted from quality control is applied to the non-suspected communities. The latter is to minimize the risk that affected communities could go undetected.

The procedure to follow during fieldwork is simple. However, the logic behind it is less than straightforward. Depending on local conditions and severity of the problem, the difficulties faced during LIS survey are as follows:

- The suspected communities are fully enumerated and the non-suspected ones form the only object, within this survey, of a sampling approach. This calls for special expertise, against the intricacies of which country surveys have to be immunized by simple rules for field-work and by outsourcing the analysis.
- Many of the survey users do not understand that there are two rather than one accuracy parameters; in other words, that it cannot

absolutely certify that no more than x percent of the non-suspected communities are affected. The solution is to base the sample size in an affected district on a probability (10%) as well as on a confidence level (90%) that are standard for the country surveys in general. The confidence level of 90% is not optimal, but agreed upon due to limited resources.
- Against all intuition, the size of the sample does not grow proportionately with the number of non-suspected communities in a district. The sample size quickly moves towards an upper limit, which is not exceeded even in the case of large districts.

This note explains the logic of the procedure implemented in practice fieldwork defining the general rules for fieldwork. It designates some areas in which specific country surveys have to exercise discretion because of unforeseeable location specific conditions or efficiency concerns affecting the outcome of landmine impact surveys. Altogether, the aim is to reduce the risk and not to take unnecessary risks in landmine management problem.

2. Types of Suspects: Areas and Communities

Intuitively, when classifying areas of an affected country as target for preliminary landmine intelligence gathering, there are at least three levels of designation of geographic areas: areas that are definitely landmine free, areas that are probably landmine free, and areas that are probably landmine affected. Rules can be made for making these categorizations of geographic areas more accurate. For example, if an official military source, an NGO working actively in a district and the district's political head all suggest that there are no mines in the district, it might be listed in the definitely landmine-free category. If the military source and the district's political head confirm that there are no mines, but a truck driver says they heard of a person being hurt two years ago, then it might go into the probably mine-free category. Finally, if the military source reports that an area has landmines, then it might go into the probably mine-affected category. Thus, an account must be made for possibilities and different sources of information.

Once the intelligence gathering survey staff have reached down to the district headquarters and are interviewing experts there, the pre-classified categories change based on up-to-date information as it is acquired. They are no longer dealing with areas like provinces or districts; they are aiming at communities, the basic units of the survey. It may still be reasonable to set aside part of the district as definitely landmine-free. This will make sense if the conflict which resulted in the placement of landmines never reached that part of the selected communities.

In conclusion, the Landmine Impact Survey comes into contact with communities for different methodological reasons associated with subsequent mine clearance activities. One methodology designated by outside experts and subsequently surveyed by field teams is only one of them. Logically, the various groups of interest are defined by all possible combinations of these properties: (suspected/not suspected) x (visited/not visited) x (surveyed/not surveyed) x (sampled/not sampled) x (searched/not searched) components. However, such a scheme will only confuse the practitioner, quite apart from the fact that many of these combinations are not possible under the rules of the protocol. However, these key properties provide an approximate picture of the real situation and are found to be useful in judgment of the extent of actions involving suspicion, visiting, surveying, sampling and searching moments of the mine impact survey activity.

Most of the communities in which survey staff visit will fall into one or more of the six categories listed below:

- Suspected, found affected and surveyed (true positives).
- Suspected and found to be mine-free (false positives).
- Sampled and found to be mine-free (true negatives).
- Sampled and found to be affected, surveyed (false negatives).
- Visited during search around false negative, found to be mine-free (true negative).
- Visited during search around false negative, found to be affected, surveyed (false negative).

Other categories include the large group of communities which were neither suspected nor sampled, and which therefore will not normally be visited. However, some of these communities may be visited in order to collect expert opinion on the place. Some of them may in fact be found to be mine-affected. They also include communities that were either suspected or sampled, but were not visited or, if affected, were not surveyed. These latter are anomalies and will normally be few in number. If their number is considerable, they reflect poorly on the quality of the landmine impact survey to be conducted.

While the expert opinion on affected communities may not always follow an easily discernible pattern, the communities visited for the control of false negatives display a clear structure. They form a systematic sample from the set of non-suspected communities within an affected district and call for an analytical statistician. On the one hand, this tends to pose intellectual challenges for the survey staff. On the other, it gives the survey the aura of scientific quality control and greater confidence in decisions to be made.

After almost a decade of humanitarian mine action, the global landmine problem remains poorly defined. A fair amount is known about the suspected location of minefields, but little is known about the socio-economic impact of landmines upon infected communities. Without information about this impact, it is impossible to develop effective strategies that seek to minimize the human and economic costs. If the terror of landmines is to be constrained within years, as envisioned by the Ottawa Treaty, rather than in decades as assumed in the early days of mine action programs, then better quality information is an immediate and unavoidable requirement for a successful landmines clearance mission.

3. Global Landmine Survey

The global landmine survey is also called Level One Impact Survey or Survey Level One (SL1). This survey has the following outcomes:

- Accurate information for strategic planning and resource mobilization. In particular it provides:
    - Clearly defined needs and program requirements.
    - Informational tools for planning support and reporting.
- Baseline data to measure progress and support country's Mine Action Centres by providing:
    - Clear measures of performance.
    - Well supported comprehensive quality assurance.
- A platform to support multi sector programming and research efforts.

In a unique cooperative effort, the NGO communities, in collaboration with the United Nations Mine Action Service (UNMAS), and the Geneva International Centre for Humanitarian Demining (GIC) established the Survey Working Group in May 1998. The Survey Working Group (SWG) facilitates the international coordination of resources and expert personnel for the completion of Level One Impact Surveys at different locations. This step is important in centralizing the information and coordination of activities and a better utilization of expert resources, as well as exchange of information to gain experience in methods of demining.

The Survey Working Group mandated the Vietnam Veterans of America Foundation (VVAF) to manage and to serve as fiscal agents for the survey action centre in the implementation of the Global Landmine Survey Program. The USA, with its enormous military power and engagement in production of landmines and war, certainly has a major responsibility to assist in cleaning its own remainders.

Moreover, this Survey identifies the general outline of the landmine problem at the local community level, through rapid rural appraisal techniques developed by NGOs and international institutions for general development and emergency programs. It focuses on community knowledge and perception of the social and economic impacts of landmines, and the general location of mine contaminated or suspected areas. It should be noted that trained enumerators from the local area carry out this survey with no specialized Mine Action training other than mine awareness training.

Through a controlled group interview process, the interviewers will gather information about mine victims, and suspected mined areas and their socio-economic impact. Because they lack the technical expertise to enter mined areas or assist the victims, the enumerators will terminate their inquiries at this general level. However, enumerators will undertake a visual proximity verification of the reported areas from one classified as safe.

This information is collated and entered into the Information Management System for Mine Action (IMSMA) developed by the Geneva International Centre for Humanitarian Demining for the UNMAS. IMSMA allows economic and geographic data to be integrated into Geographic Information Systems (GIS) permitting a variety of analysis and interpretation. An indexing system will provide a community risk profile based on three negative categories (victims, blockage of economic assets, blockage of infrastructure), and one positive factor, as a mine awareness program in the community. The architecture for the indexing system will be open so that national and regional groups, or groups with a particular sector interest, can modify the system to meet their individual needs based on local conditions.

The completed survey will provide data for the development of priorities and improved planning in existing programs, and on overall national priority and works plans where no such programs may yet exist. The data will also provide a baseline against which to enable the measurement of progress. The information provides policy makers at the national and international level with information at the village or community level on the social and economic impacts of landmines on lives and livelihoods of affected communities. Through this process, a ranking of problem areas can be identified geographically and then mapped, thus permitting policy makers to focus efforts in the most affected areas with the highest priority of clearance. These Surveys enable mine action specialists to specifically target programs for various interrelated activities such as training, mine awareness, mine marking, level two technical surveys, as well as clearance and victim assistance.

4. The Survey Process

The survey begins at the national level, where survey teams meet with national experts and obtain information about what areas of the country are known, or thought to be contaminated. After getting this general information, the survey teams will gather more detailed information at the local level. Data collected at the local level is then analyzed and presented in the impact survey. Local community interviews are at the heart of the overall survey process. Hand drawn maps of contaminated areas are made with the assistance of local leaders of known or suspected affected areas. In this way, it is possible to clearly point out the relationship between the affected area and key local landmarks, and valuable assets such as fields, water sources or key roads. It is then possible to identify what type of threats and blockages are created by mine and ERW contamination. In parallel, information about ERW and mine incidents is also collected. Upon completion of the community interview stage, the data collected is analyzed to determine the level of community impact.

A level one impact survey is conducted in different phases. The phases and their component activities are presented in a general chronological order, which correspond to the task/sub-task matrix of the certification guidelines. A number of activities (or tasks) will occur within each phase, although in some cases a given task may continue throughout two or more phases. This general process may be modified as required by the particular conditions encountered in any given country, particularly if that country has a pre-existing mine action program which can serve as a roadmap to create new and modified programs.

A standard mine action program has four phases. Each phase is described details as follows (Kievelitz et al. 2004):

Phase 1: Planning, preparation, and office establishment

UN Assessment Mission*:* The United Nations Mine Action Service (UNMAS) is conducting a series of assessments in mine affected countries to determine the nature of the problem, and affirm the host nation's support for humanitarian Mine Action. The recommendations provided by UNMAS assessment missions indicate the need for a level one impact survey. UNMAS begins the process to promote the establishment of a sustainable national mine action program.

A UN assessment mission is an expected precursor to humanitarian mine action intervention for Mine/UXO affected countries. The assessment mission will first examine the "need" and "utility" of mine action within a country's existing institutional mechanisms. This mission explicitly calls for the execution of a level one impact survey when it is justified. An assessment mission report calling for the conduct of a survey is the first step in the process.

Advance Survey Mission: UNMAS formally requests that the survey action centre mobilize an advance survey mission to the identified mine-affected country. The advance survey mission establishes liaison with national authorities, in-country NGOs and the UN to develop the working relationships needed to execute a level one impact mine / UXO survey. In addition to confirming the "need" and "utility" for a survey, this mission will also examine the "feasibility" for the actual execution of the actual landmine survey.

The purpose of an advance survey mission is to develop contacts and information required to plan for the actual survey. This mission will coordinate closely with national authorities, and as an outcome critical planning information is gathered and a capacity assessment made of organizations and resources, which may be capable of supporting a survey. Using this collected information; the survey action centre prepares a country report/proposal and develops the initial survey operations plan. In some cases, information may suggest that a survey is not appropriate and as a result all activities will stop, or be delayed.

Operational Set up and Establishment of Country Presence: Upon the approval and funding of the impact survey, an operational base(s) will be established to support the survey team, and it will further be completed with temporary offices and a communications system.

Transparent recruitment and selection process is to be undertaken to attract and retain competent local and international staff. Ideally, each country survey team will include members with a wide range of skills including social science methodology, information management and technical mine expertise. The final selection of staff is the responsibility of the implementing survey team with due consideration of the national host requirements.

Collection of Expert Opinion: The survey team will begin the systematic collection of informed opinion of national "experts" familiar with the mine contamination problem. This should ideally be done by using a standardized interview format, and should involve a broad cross-section of agencies, and individuals representing diverse interests and regions. This interview process should build on the contacts established during the advance mission and should produce a basic understanding of areas within the country suspected of having mine/UXO contamination. This is an ongoing activity and will be repeated as needed throughout the survey, especially as the actual enumerator team move into new areas or regions. The expert opinion could also be alternatively collected through the mailing of forms to various organizations across a country.

Sounding Board Meeting: The global survey instrument and methodology will be discussed with local stakeholders including relevant government bodies, NGOs involved in mine action activities, and experienced social scientists to ensure that it meets local requirements and accounts for cultural sensitivities. This could be done through extensive consultation with relevant local experts and holding of a workshop to fine tune the survey instrument and methodology to meet local needs and requirements.

Phase 2: Project expansion and survey instrument refinement

Training: Based on the selection process and the level of academic and professional skills of the survey staff recruited, a training program is to be developed. The aim of the training program is to teach survey related skills and knowledge to smoothly reinforce the survey process. The team training should be based on the expected outputs and requirements of the survey mission.

Conduct Pre- and Pilot Tests: Following the survey training stage, a pre-test of the survey instrument itself is to be conducted to determine its utility and cultural acceptability. Likewise, a pilot-test is to be conducted of the entire survey process to analyze management, logistics and administrative systems and planning assumptions. This is a critical stage in the survey process; decisions made at this juncture must be well documented to support the full survey process and implementation. A complete review and analysis of the pilot test results and adjustment of the survey process is to be completed before finalizing the national level one impact mine/UXO survey implementation plan.

Develop Survey Sampling Methodologies: The refined survey process is to be used to develop methodologies for conducting both a sample survey as well as a sampling to check for prevalence of the false negative community category. These will be used to build a predictive model of the survey results as well as to build a measure of confidence regarding survey findings.

Revise Operational Plan: Prior to actually conducting the survey, information gained during the testing exercises is to be used actively to revise and update the final operations plan. This will be done in the host country and will involve input from the relevant national authorities, as well as the Survey Implementation Organization personnel, the UN and other stakeholders.

Phase 3: Data collection, processing and verification

Administer and Manage Level One Impact Mine/UXO Survey: The conduct of the level one impact survey is to take place under contractual agreements and in close collaboration with national authorities. An independent quality assurance monitor will evaluate the survey. The survey is to be conducted in accordance with the survey working group and UN policies. There will be a results-based management system established to track team assignments and outputs, rigorous field supervision and extensive quality control on data editing and entry.

Community Group Interviews: The field supervisors would visit mine affected areas prior to the survey teams to make overall arrangements for the selected communities to be surveyed by the survey teams. The teams will conduct a group interview to the key informants within a community. The questionnaire will be administered to register the information of the group, and the group interview conducted based on participatory approaches.

Visual Verification of Mined Areas: Reported mined areas will be visually verified from a safe area at a distance. Strict procedures are in place to ensure that enumerators do not take unnecessary risks. Visual verification will not only improve the quality of the data but will also provide relevant data for subsequent technical (level two) survey.

Supervision and Field Editing: Field operations will be regularly monitored and supervised. Trained local supervisors, appointed for specific areas/regions, will ensure that quality information is gathered. International staff will also conduct regular field visits. All data will be checked and edited by a full time data editor, as close to the field as possible, and prior to this they are entered in the computerized database.

Quality Assurance Monitoring: Survey quality is assured by a quality assurance monitor who records and documents survey progress in accordance with the UNMAS Certification Guidelines. To ensure the high quality of the survey process and results, an independent quality assurance monitor (QAM) team will be tasked by the funding/contracting agency to monitor and document all critical activities of the survey for corrective actions. The QAM's reports will also serve as an instrument for certification of the survey results.

Phase 4: Data analysis, presentation and hand over

Data Analysis: The objective analysis of data collected during the survey is vital to ensure acceptance of the survey results. The analysis of collected facts in quantitative study and impact, taking into account that if the study is applied it should be applied in a transparent, logical, and understandable way. The analysis of data must meet the informational needs of the national authorities.

Handover: As part of completing the survey, the Survey team is to present its preliminary findings, maps and data to the national authorities and other relevant stakeholders in the affected country. This is not only to share the knowledge gained, but also to elicit comments and input prior to publishing the final report and to releasing data on the internet to be available to the research community, and for the purpose of comparative studies across communities and countries.

Prior to departing the country, the survey team is to provide all information to be followed by the local mine action agency. This information is extensive and will include a package of data, maps and a draft written report.

However, there are systematic shortfalls for this system, the major limitation being that as the survey reports the areas that are affected or believed to be affected, the survey assesses perceived risk, and at this stage of the process areas reported is not verified and mapped.

Therefore, the area reported as affected may be larger than the actual area so the survey inevitably presents a snapshot of the problem. This can provide a great deal of information and bring previously unknown or under-appreciated problems (like the presence of ERW) to light. However, this survey cannot track a fluid situation. Changes brought about by clearance activities, a resumption of conflict, the discovery of previously unsuspected affected areas, or natural causes (floods and earthquakes can displace ERW and mines) will not be reflected in the survey.

Reports: The survey team may produce a final report as a product for the national authorities in order to allow them to continue the ongoing mine action process. This report is to focus on an objective presentation and analysis of the data gathered in the survey. The survey team leader will also produce a more subjective report providing commentary and recommendations for how future mine action activities are to be undertaken.

5. Summary of Landmine Impact Survey

Landmine impact surveys are grown out of the worldwide collaborative efforts of the international campaign to ban landmines and the Ottawa treaty. They are executed to meet the overlapping needs of international donors, national authorities in landmines and UXO affected countries as well as humanitarian mine action planners and implementers.

The overall vision of landmine impact surveys as articulated by the Survey Working Group (SWG) is to "facilitate the prioritizing of human, material, and financial resources supporting humanitarian mine action at the national, regional, and global levels." To fulfil this vision, landmine impact survey projects are executed to meet the same high and robust global standard. Their implementation facilitates an improved collaboration between the three primary stakeholders of the process: national authorities, donors and implementing agencies, and in addition provides for the utilization of the same dataset for other purposes such as research.

The data collected during the survey process provides clear improvements on past efforts as follows:

- Defines the entire problem in terms of scale, type of location, hazards, and socio-economic impacts upon communities.
- Improves national planning efforts by allowing for a better prioritization of resources.
- Fosters development of national plans with well-defined immediate, intermediate and long–term objectives.
- Establishes baseline data from which outcomes and performance can be measured.

The landmine impact survey is considered to be the first and most vital step in the transformation of overall humanitarian mine action (HMA). These surveys are intended to dramatically improve the quality of information available for decision makers at all involved organizations and levels.

Subsequent meetings of the SWG further refined this vision to create a number of subordinate objectives focused upon three key constituencies namely to:

- Allow donors to rationalize their allocation of funds to the areas of greatest human need, as defined by the impact of landmines and UXO upon communities.

- Permit national authorities to develop national plans focusing upon regions and areas of greatest impact.
- Give implementers baseline impact data that will provide success indicators for mine action programs.

As a global initiative to standardize available information regarding the landmine problem around the globe, the landmine impact survey process represents a successful and concerted effort to ensure conformity of methods, procedures and processes. These surveys are based upon the best practices in the fields of social science research and humanitarian mine action. To ensure quality of results, internal and external quality assurance mechanisms are imbedded into the methodology of the survey process. Landmine impact surveys are executed to internationally recognized standards and measure and score impacts upon affected communities in a generally uniform manner.

In addition, due to the frequent presence of mine threat worldwide and low capacities for clearing at international related institutions, the affected countries are expected to allocate internal resources to prioritize the local activities of mine clearance. As the international and NGO bases resources are limited, the available resources could be strengthened by taxing the mine producer countries. The national priority activity can be achieved through conducting a general landmine impact survey. The survey will supply the necessary information about the magnitude of the problem and resources required for clearance.

The landmine impact survey assesses the impact of explosive contamination on nearby communities. The main purpose of conducting the survey is to assess the mine situation in each mine and ERW affected community, and to develop an understanding of how this threat affects the related communities. The primary objective of the survey is to collect useful data about mines and ERW and in addition to support further steps in the mine action process including marking, clearance and completion survey phases.

In this chapter, a number of issues were discussed in detailed form; the purpose of the landmine impact survey, the types of suspects including areas and communities, the issue of global landmine survey and the actual survey process, with each component of the survey process discussed. In sum, the chapter provides a complete set of information about the organization of the landmine impact survey and its implementation and outcomes. The information is vital to the formation and function of national capacity to clear landmines and further issues of settlement of legal issues.

Notes

[1] A study of the role of survey in mine action (2006), GICHD, Geneva, Available at: http://www.gichd.ch/fileadmin/pdf/publications/Survey_in_MA_March2006.pdf, Accessed on Dec. 12, 2009.

CHAPTER SIX

LANDMINES IMPACT SURVEY IN THE KURDISTAN REGION

This chapter provides an introduction to the mine action program in the Kurdistan Region. It serves as an overview of the mine action activity in the region by focussing on the basic original approach and how it was developed through the use of an integrated approach and the differences in their outcomes. The background, implementation, outcomes and lessons learned from the integrated approach are discussed in a more detailed form. It presents detailed statistics about the mine action resources, local capacity building, fields and classified communities and categorized victims by their gender, age and profession. A summary of the findings and its descriptive analysis are provided.

1. Mine Action in the Kurdistan Region: A General Overview

The mine action conditions in the Kurdistan region comprises all the problems associated with mines contamination encountered in the previous chapters. In general, it focuses on transfer and use of mines, mine awareness education programs, mine victims assistance, survey of mine-suspected areas and mine clearance activities.

In the Kurdistan region, the mine action program is implemented through three interrelated and complementary components as follows:

- Minefield surveys, mapping, marking, and demining.
- Mine awareness and risk reduction.
- Mine victim assistance, including rehabilitation and reintegration.

These basic activities incorporate the main elements of dealing with mine problems. In addition to these basic activities, there are numerous supporting activities which are also crucial for the success of mine action, such as: setting up standards, carrying out necessary evaluation studies, resource mobilization, management development and skill upgrading, integration of science into solving problems of mine action by involving researchers at universities in the different actions activities, etc.

The basic task of the Iraqi Kurdistan Mine Action Centre (IKMAC) is to plan and implement mine clearance operations in the region. In order to achieve the main task, numerous related activities are required, such as:

- Gathering and processing of data regarding different mine suspected areas, mine contaminated areas, mine cleared areas and buildings, areas excluded from the category of mine suspected areas, mines and ERW that have been found and destroyed, and other technical characteristics of mine suspected areas. The required documentation is being stored, and the database is being maintained, processed, and managed through the use of IMSMA.[1]
- Conducting general and technical surveys of mine-suspected areas.
- Cooperating with local authorities (development of the county's demining plans and minefield density).
- Planning for demining operations (develop project proposals and scope of work).
- Creating documentation for public tenders.
- Implementing competitive bidding for demining projects and the pyrotechnic search of areas or buildings in accordance to the plan.
- Performing monitoring and quality assurance of demining activities.
- Performing quality control and issuing certificates of mine clearance.
- Coordinating mine awareness education.
- Coordinating assistance to mine victims.
- Cooperation with international entities participating in demining.
- Marking of mine suspected areas.
- Training of demining personnel and assisting staff.

Moreover, to fulfil the abovementioned national tasks, numerous supporting tasks and human skill development and organization structure and capability enhancing activities were performed. For example, training employees, training different demining experts, team leaders, equipment operators, computer technicians, finances and budget experts.

2. The Integrated Approach Program

2.1 Background

As described above, the mine action program in the Kurdistan region consists of three core components: mine clearance, mine risk education (MRE) and mine victim assistance (VA). These core activities will plan their work ahead in isolation from each other resulting in a non-synergetic approach to demining. Given the in-optimal resource utilization and outcome, the United Nations Office for Project Service (UNOPS) has implemented a new integrated approach program across all three core activities within its strategic plan of 2003. Subsequently, all UNOPS mine action activities have been pre-planned in annual work plans considering community priority and size of mine-affected areas resulting in a cluster tasking approach.

In accordance with the changed policy and approach, the impact of landmines on communities includes different factors. The most important factors as described by Matthew et al. (2004) are as follows:

- Blockage of economic sources.
- Blockage of housing.
- Continuous accidents leading to threat to people lives.
- Blockage of access to educational, religious, and historical locations.
- Ignorance of how to deal with mines and minefields.

All these factors in various forms and degrees impeding the community's livelihood have been addressed so that villagers in these affected communities can resume their normal day to day activities without being subjected to risk.

The aim of the integrated program was to coordinate all of the UNOPS mine action programs (MAPs) three core activities (i.e. clearance, MRE and VA) in order to address all impact factors on each individual community with the objective to relieving the community from the impact of landmines and ERW.

2.2 The Program Implementation

In order to implement the integrated approach program, UNOPS-MAP teams were made up of mine clearance (MC) specialists, mine risk education (MRE) specialists and victim assistance (VA) specialists who worked together with villagers to ensure the reduction of landmine impact.

Mine clearance specialists conducted technical surveys to assess the need for clearance and marking activities. Mine risk education specialists solicited the support of villagers, kept them informed on the work-plan progress, provided safety information, addressed issues arising from clearance activities, and evaluated the post-clearance impact on the individual communities while victim assistance specialists assessed the need of victims in need of medical, surgical, orthopaedic rehabilitation, and vocational training services.

In this holistic approach, an integrated plan for clearance, victim assistance and mine risk education was developed in collaboration with local services, contractors and the mine coordination office (MCO). UNOPS Mine Action Information System (MAIS) supported the implementation plan through collection, management and supply of vital input and output of data. Moreover, the implementation process consisted of four stages including planning, tasking, execution, and analysis of results. Fig. 6.1 below illustrates the implementation of the integrated approach process.

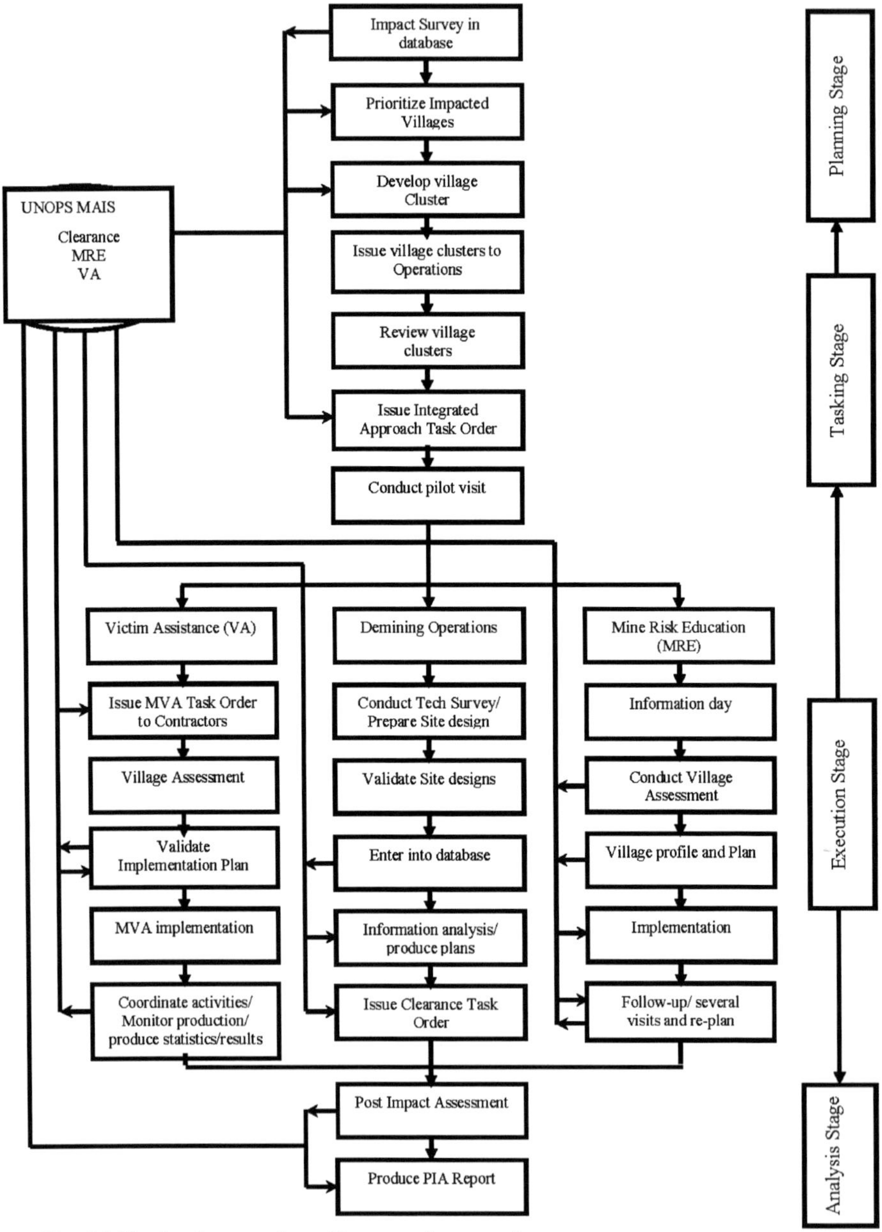

Fig. 6.1 The implementation of integrated approach process

2.3 Lessons Learned

The introduction and implementation of the integrated approach described above has led to a change in work procedure and outcome. The following points are a summary of the main lessons learned within the integrated program:

- There is evidence that coordination and cooperation between the core activities of UNOPS-MAP made strong impressions and impacts on communities.
- Planning, coordination and cooperation with local authorities/ MCOs addressed the impact effectively.
- Reducing visits to villages by consolidating key personnel with "Pilot Visits."
- Through providing MRE to the community, villagers understand clearly the difficult nature of mine clearance and the reasons for the slow demining processes.
- The integrated program highlights the interest and cooperation of the community aimed at alleviating the impact of landmines.
- The integrated program illustrates the level of coordination to effectively deliver critical services to priority mine-affected communities.
- The integrated program fosters a better understanding of the whole UNOPS-MAP amongst the staff of each component, especially the objectives, vision and mission of MAP.
- The new approach has led to identifying many victims who belong to the category of non-war handicapped people found in the villages, and that no national or UN plan is in place to address their needs.[2]

2.4 The Power Line Survey

The UNOPS MAP as the primary contractor for mine action operations in the Kurdistan region provided demining support to the United Nations Development Program UNDP Electrical Network Rehabilitation Project (ENRP). Support to UNDP/ENRP commenced in February 2002. The ENRP aimed to improve the electricity network in the region by rehabilitating the existing and extension of new power transmission lines. The total length of the power line grid is approximately 2,700 km. UNOPS-MAP conducted a survey of the grid and identified that approximately 13 km of these lines had minefields either directly under or in the immediate vicinity (up to 25 metres from either side of the towers).

2.5 Verification of Minefields Cleared by Others

This process was launched in August 2001 to verify all minefields cleared by the Mine Advisory Group (MAG), the Norwegian Peoples Aid (NPA), Solidarity, villagers, local authorities and other relevant parties in the Kurdistan region. The information collection process was completed in December 2002. It should be noted that due to limited resources, the clearance of these minefields was not conducted in accordance to norms and procedures that have been established by the IMAS.

In order to confirm the clearance of these minefields, UNOPS-MAP embarked on a procedure that involved the Mine Coordination Office (MCO) with the impact survey teams to use a legal agreement document produced by the UN. The document explained the non-liability of UNOPS-MAP of any accident that might happen in the areas verified by the UNOPS Survey Teams.

2.6 The Post Clearance Impact Assessment

The post clearance impact assessment (PCIA) process was introduced to address any cleared minefields by the UNOPS Mine Action Program through an assessment after a six-month period of clearance.

The UNOPS head office plans section were to analyze each cleared minefield in the database and issue a list of handed-over minefields to the sectors. The national operations assistant, the national survey teams, and location supervisors were accordingly tasked to provide information requested in the post clearance impact assessment form and returned to the head office plans section. The MRE and the VA representatives at the sector level also gathered relevant village impact data for the final PCIA report.

2.7 Conclusion of the Integrated Approach Program

The UNOPS Mine Action program (UNOPS-MAP) revised its priorities of independent minefields to focus on a community priority and size of mine-affected area, resulting in the transfer to a cluster tasking approach.

In addition, the UN agencies and local authorities were involved with determining suitable weights for different village blockage factors. This radical change in the process yields a better indication for identifying the needs that are the greatest within the Kurdistan region. The UNOPS-MAP three core resources (clearance, MRE and VA) can focus on addressing these specific requirements. This holistic approach provided an integrated response, designed to address community needs and alleviate the blockage factors, thus having a greatest positive impact on the people and local communities.

The integrated approach spanned the UNOPS core functions and attempted to synchronize their previously non-integrated efforts. This approach was launched on August 15, 2001. Moreover, the United Nations Office of the Iraq Program (UNOIP), which managed the "Oil for Food Program," was established under the Security Council Resolution 986. From 1997 to 2003, UNOPS managed the integrated Mine Action Program (MAP) in the Kurdistan region on behalf of the UNOIP.

Under this integrated MAP, more than 12.2 million square metres of high priority landmine and ERW contaminated land was cleared. During these clearances, the UNOPS-MAP identified and destroyed over 79,000 UXOs and ERW, 2,500 cluster bombs, 11,000 AP mines and 560 AT mines. In addition, more than 7.9 million square metres of cleared land was handed over to the local population for productive use and a further 24 million square metres of contaminated land was marked. The program also has provided MRE to more than 290,000 people and 380,813 treatment and rehabilitation services to mine/war victims through a network of two hospitals,[3] eight health clinics and another twenty First Aid posts.

As a result of the previously mentioned activities, through a combination of marking, clearance and MRE, the landmine victim incidence rate was reduced by over 50% between 1999 and the end of 2002. UNOPS-MAP focused on growth in its activities via a national capacity development program in order to address the landmine and ERW contamination problem and provide long-term sustainability in the effective demining of the region.

This involved establishing four local and manual demining NGOs, two MRE NGOs, three local Mechanical contractors and development of a local capacity to plan and manage the full spectrum of mine action activities. This was a major achievement, as the original MAP mandate was simply emergency relief oriented. Transfer of all manual teams took place by mid-2003 and all dog teams were transferred in October 2003. The MRE and the MVA components were entirely managed by national entities since 2002 (UNOPS Mine Action Unit).

3. The Landmines Impact Survey

3.1 A General Overview

The landmine impact survey (LIS) in the Kurdistan region was conducted by UNOPS-MAP in the late of 2002. It was then followed by a landmine impact survey for Iraq in general (including the Kurdistan region), which was conducted over a two years period and completed by late 2006. [4] The LIS in 2006 was implemented by the Information Management and Mine Action Programs (iMMAP) that was previously a part of the Vietnam Veterans of America Foundation (VVAF).

The LIS implemented by UNOPS confirmed and complemented the original general survey conducted within the region. The general survey was conducted the periods of 1998 to 2000, building in turn on data collected during the previous years by Mines Advisory Group (MAG). During the process, various teams have visited all communities in the region and identified 3,444 minefields and 1,096 mine-affected villages. While the LIS implemented by iMMAP visited 4,291 villages and identified 1,126 landmine and ERW affected communities, [5] with a population of 748,651 at risk due to 3,024 landmines in suspected hazardous areas with a total size of 776 square kilometres found during the survey. The two comprehensive coverage surveys have served to complement each other and provide a detailed picture of the magnitude of the mine problems in the region.

The two different periods of the LIS were conducted to identify specific blockages to the community's social and economic activities created by the minefields, and then to confirm the information previously collected, and is thus an effort to retrospectively fit a comprehensive minefield database with the socio-economic blockage information of an impact survey. A total of 1,126 mine affected communities were surveyed during the landmine impact survey. However, 39 minefields associated with some affected communities could not be surveyed for blockage factors due to security restrictions based on the general survey information. These minefields have a total area of approximately 60 square kilometres (see Table 6.1 below).

Table 6.1. Landmine Impact Survey Effort

Total Communities Visited	4,291
LIS surveys conducted	1,126
Un-surveyed communities which share affected areas with surveyed communities documented.	234
Communities with Only ERW Spot Reports.	67
Total Communities with Any Kind of Contamination /Impact.	1,427
Total ERW spot reports documented.	289
Total communities documented not contaminated.	2,864
Total communities documented previously contaminated.	541
Total abandoned communities documented.	461
Total new communities found.	336

The distribution of the landmine and ERW affected communities classification is indicated in Table 6.2. In regards to the impact score, in alignment with initiatives of the Survey Working Group (SWG), the weights that assigned for the blockage factors have been applied to the region and then normalized for ease of comparison to other threats. In the LIS report, impact scores ranged between 1-5 represent a low impact, 6-10 a medium impact, and greater than or equal to 11 represent a high impact community.

Table 6.2. Distribution of impact categories of communities

Score Interval	Community Classification
1–5	Low
6–10	Medium
11 and higher	High

Twenty-five to twenty-seven districts within the Kurdistan region are mine affected, in which 97 to 124 sub-districts and 1,126 of 4,281 communities (villages) are included, with an estimated total of 748,651 persons living in the impacted communities, giving a total population of 3,547,000 people in the region.[6] The figures suggest that more than one in every five people is living in a mine affected community.

According to the latest LIS (iMMAP-LIS), the governorate of Suleimaniya contains the majority of mine-affected communities (582 out of 1,126) with a mine affected population of 266,559 people. There is a high concentration of mine affected communities in Choman (located in the Erbil governorate), and in areas such as Halabja and Penjwin (located in the governorate of Suleimaniya), but no high impact communities have been found in the governorate of Duhok (see Table 6.3 below for summary of the high, medium and low impact communities by governorate).

Table 6.3. Landmine ranked Impact communities in Kurdistan Region

Governorate	High Impact Communities	Medium Impact Communities	Low Impact Communities	Total
Duhok	0	70	175	245
Erbil	4	120	175	299
Suleimaniya	24	254	304	582
Total	**28**	**444**	**654**	**1,126**

Moreover, the LIS report identified a total of 3,024 distinct areas suspected to be mines and/or ERW contaminated, which were assessed with respect to different blockage factors related to economic sources, housing, continuous accidents and threats, educational, religious and historical locations, and ignorance of risks.

The key informants in the affected communities reported that the suspected areas cover a total land surface of 776 square kilometres. The mean size of an affected community is 112,746 square metres. It should be noted that 55.6% of the quantity of mined areas are between 10,000 and 100,000 square metres. A total of 23.1% of the communities are larger than 100,000 square metres, and 14% of them are smaller than 10,000 square metres. The remaining 7.3% of the surveyed mined areas have unknown dimensions because they are located either within restricted areas or suspected mined areas cleared by other parties requiring verification.

The large scale of contaminated areas and the dense population has resulted in large number of accidents. The distribution of the (474) recent victims[7] by age and sex indicates the characteristics of the groups of concern. For example, 94.6% of the victims were male. Men ranging in age from fifteen to twenty-nine account for 44.8% of all victims compared to women of the same age at 1%. The most frequent categories classified by occupation involving accidents were: children playing, shepherds, military, farmers, and workers, accounting for 87% of all recent victims. The type of activity at the time of accident was closely related to the occupational breakdown as above, with the notable exception of nearly 30.8% of the recent victims being involved in or close to someone tampering with mines and ERW. Tampering with munitions is the most prominent cause of accidents for males and second most prominent for females. The pattern is clearly indicating a need to pay more attention on mine risk education and EOD activities.

Table 6.4 indicates the distribution of the total affected areas by ordnance, types, and governorate locations. The statistics show that total sum of 2,946 mine affected communities are representing communities containing AP landmines, while another 97 communities representing those communities that are containing AV landmines. These two ordnance types are the main types across the three governorates.

Table 6.4. Distribution of Ordnance, Type and Governorate

Ordnance Type	Erbil	Duhok	Suleimaniya	Total	Percentage
AP Landmines	646	698	1,602	2,946	97.4
ERW / UXO	190	117	749	1,056	34.9
Cluster Sub Munitions	30	24	7	61	2.0
AT/AV Landmines	26	14	57	97	3.2
Stockpiles	14	4	1	19	0.6
Other	8	1	3	12	0.4
Total affected area	679	709	1,636	3,024	100.0

The types and distribution of munitions in contaminated areas also affect the choice of employed clearance techniques. The survey sought information about AP landmines, AT landmines, cluster sub munitions, other ERW/UXOs and any stockpiles that might have existed. These are somewhat generic categories and the survey did not seek to provide more specific information on the landmines (such as specific type of AP landmine).

It should be noted that since a mined area may be contaminated by more than one type of munitions, the frequencies reported in Table 6.4 above will add up to more than the total number of affected areas.

3.2 Clustering

As part of the integrated approach, and the priority criteria, a clustering system was introduced into the UNOPS-MAP in 2002. In the design of the clustering system, UNOPS-MAP was concentrated on a group of mine affected villages. The head office (HO) planning section was responsible for developing village clusters based on the classified mine affected villages ranked as high, medium and low. These classified villages were issued to the geographical information system section in the head office

HO-GIS to plot them on maps and accordingly, HO Plans to work on those maps and develop clusters manually.

Villages with minimum minefields and small areas had the first priority. Therefore, clusters should contain these small villages as much as possible. In many cases villages with large areas of minefields were included in the clustering due to their geographical vicinity to each other and to afford a sufficient work plan for the demining activities to sectors. The procedure and the organization of activities by clustering is to economize on scarce resources and their effective utilization in a mission that is vital to each community.

3.3 Blockage Access

The three activities mainly denied land in the region were non-agriculture uses, pasture, and agriculture areas, accounting for about 79%, 98%, and 54% shares of the respective land area. Pasture areas (migratory pasture and fixed pasture) are the most affected land category by landmines, particularly around strategic hillsides and mountain tops that were former military strategic positions. Given the nomadic nature of the shepherds and their prominence throughout the region, this figure is not surprising and suggests one out of every three mine impacted communities is strongly affected by landmines impeding grazing land areas.

A number of 1,103 communities were affected by blocked pasture land, affecting some 906,469 people and representing 1,475 mined areas or a total of 785.3 square kilometres. This grazing land represents more than two thirds of the estimated mined area in the region.

In conclusion, with the completion of the landmines impact survey in the Kurdistan Region, the mine action community collected a comprehensive set of mine related socio-economic impact data at its disposal. This data allowed the mine action community to develop effective plans to target areas that are posing the greatest threat, and to identify communities bearing the greatest impact. The major findings are summarized as follows:

- The mix of high, medium, and low mine affected communities in close geographic proximity is suggested by a distributed clustering approach to task clearance assets, not necessarily targeting only high impact areas first. This approach recognized the size(s) of the mined areas to maximize the greatest effect in the shortest term.
- The majority of mine-affected communities were villages, and the majority of mine impacted activities were of a rural nature. The

long-term response to the landmine problem must focus on the majority of resources in the rural areas.

- The collective towns and district centres received a high priority for clearance, and the impact of landmines on them can be fully relieved in a single season using only a minority or small fraction of MAP assets.
- Based on the new policy, there are districts with total contaminated areas of under one square metre, for which the impact could be totally eliminated in a single clearance season with a minority of MAP resources.
- Mine affected communities suffering no socio-economic impacts have been targeted for early minefield marking rather than focusing on clearance (1,322 minefields in 572 mine-affected communities). Some communities with no socio-economic impact at the time of the LIS have experienced significant impacts with the changes in movement of the population and land use that could develop in the post oil for food program (UN resolution 986 in 1996) situation.
- Minefields blocking access to water sources have showed a positive correlation with the number of victims, and received priority targets for clearance. Clearance in such cases may often involve establishing a wide access breach through the minefield, with marking of the remaining areas.
- Minefields around community public facilities (schools, clinics, religious sites) should be addressed on a priority basis, whether to undertake first clearance or simply just marking.
- Minefields blocking access for construction of major infrastructure for the region (for example electrical power grids) have needs to be addressed as a priority issue in this respect.
- About 20% of recent victims of the last two years of accidents sought medical treatment, while roughly 80% suffered from ERW accidents. This suggests the importance of increasing focus on ERW in MRE and clearance.
- The MRE recommended activities targeting children and infants as one critical group. Since nearly half of the accidents result from tampering, with either mines or ERW, this clearly indicates a special focus area for mine risk education teams.
- One sixth of patients to the Emergency Hospital[8] system had an injury or accident relating to mines and ERW. About 25% of the victims were permanently handicapped with 14% remaining unemployed. This reinforced the requirement for Victim Assistance to sustain rehabilitation and vocational training.

With a strategically focused effort in the Kurdistan Region, the impact of the landmines can be dramatically reduced and controlled more rapidly and efficiently, providing the greatest impact in each year for the Mine Action Program and a positive outcome to the contaminated communities.

3.4 Impact on Sectors

A comprehensive general survey was conducted between October 1997 and December 2000 before launching the LIS in the Kurdistan region, the survey has identified and classified a number of 1,096 mine-affected communities containing about 3,444 minefields. Hereafter, the landmine impact survey was initiated in February 2001 for the first period and later on the next LIS in 2006, with the objective to collect intensive information regarding the types of activities for the local populations facing the threat of landmines and ERW. An impact survey is conducted for a total of 1,126 mine-affected communities that contain a number of 3,248 minefields. These figures form the basis for analysis of results obtained from the LIS.

The activities denied due to the threat of landmines and ERW were termed as "blockage factors." The most frequent impediment to villagers was the denial of access to grazing land, with nearly 78% of all villages reporting access to such lands being blocked and it accounted for approximately 30% of the number of incidents. Most of the victims recorded in 1999 and 2000 for the first LIS period and 2004 to 2005 for the second period of the LIS (representing a more complete victim dataset) were attributed to land blocked and associated with production of crops.

3.5 Implications for Mine Action

The results obtained from the LIS indicated that all the three governorates of the region suffer from various adverse consequences of landmines and ERW contamination. Although these impacts are felt throughout the region, the existence of dense "clusters" of highly and moderately impacted villages suggested the need for a well-planned and targeted set of mine action initiatives that could bring about considerable improvements in a relatively short period of time.

3.6 Communities Affected by Mines and ERW

The landmine surveys and mine clearance operations are funded by the United Nations Oil for Food Program and thus have paved the way for thousands of Kurds in the three northern governorates to re-establish their farming and grazing lands and to restore vital infrastructure to their economic life and progress.

The survey teams managed by the United Nations Office for Project Services (UNOPS), and later iMMAP-LIS later, they found that one in every four people in the three governorates of the Kurdistan region live in communities with a risk of death or injury from unexploded mines and other ordnances. However, the concentration and threat differ from one location to another.

The data collected by the two surveys was used to guide strategic demining operations and mine risk education, confirming the following:

- Landmine accidents declined.
- The community safety and confidence to pursue normal activities in cleared areas were significantly restored.
- Crop and livestock production in the cleared areas subsequently increased.

The latter survey (iMMAP survey) found that a total of 776 square kilometres in the region contained landmines and ERW. Table 6.5 below provides information about the magnitude of mine blockage access concerning the areas, communities affected, and their percentage shares of the total. Pasture land is a key factor where migratory pasture is dominant. The second highest area is non-agricultural uses of land followed by crops. The degree of contamination is highest amongst the non-agricultural uses of land type. Water usage and roads are also among other areas that show significant degrees of contamination.

Furthermore, since 1998, UNOPS worked with local demining personnel trained in mine clearance techniques, including the use of detectors, mechanical flailing machines, and mine sniffing dogs. The local demining NGOs were established in mid-2002 to ensure sustainable capacity building to localize and manage the landmine problem in the region.

In the period between 1998 and 2002, the UNOPS Mine Action Program has removed about 76,500 mines and UXOs from 9.1 million square metres of land, of which 3.95 million square metres were returned to the local population for productive use. Elsewhere, communities remained restricted in their ability to move safely from one place to another in farming their land and tending livestock, as well as collecting fuel, wood and drinking water for their home in safe way. In recent years improved living standards and the availability of energy and increased environmental awareness has markedly reduced wood usage.

Table 6.5. Mines blockage access[9]

Area of Blocked Access	Communities Affected	Percentage
Any Kind of Pasture	1,103	98.0
Migratory Pasture	748	66.4
Fixed Pasture	621	55.2
Crop Land	609	54.1
Rain-fed	467	41.5
Irrigated Crop	229	20.3
Non-Agricultural Uses of Land	888	78.9
Wild Food	798	71.0
Firewood/Charcoal	551	49.0
Wild Herbs	79	7.0
Building Materials	22	2.0
Other	254	23.0
Any Water Usage	144	12.8
Drinking Water	94	8.3
Other Water than Drinking	141	12.5
Any type of road	131	11.6
Roads to Administrative Centres	6	0.5
Other Roads	128	11.4
Infrastructure	20	1.8
Housing	9	0.8
Other Infrastructure	11	1.0
Total Communities	**1,126**	**100.0**

In addition to clearance operations, the program worked with 2,000 mine accident and war victims, providing minor and major surgery, prosthetics for lost limbs, and other rehabilitation services. Tens of thousands of women and children have received MRE in ways to navigate their environment more safely and also to report but not to touch unfamiliar objects. Mined areas that were identified but not cleared were marked with warning signs.

The UN Office of the Iraq Program invested about $110 million in the mine action program for the Kurdistan region between the years 1998 and early 2003. The resources used have resulted in the LIS identifying mine-affected communities in all three governorates in the Kurdistan Region. In total, an amount of 1,022 communities in 24 districts were found and surveyed to have mines and/or ERW problems, while 3,304 communities were found to be mine-free. An estimated 737,250 people were found to be living in those affected communities.

Table 6.6. Affected districts, communities, and population by governorate[10]

Governorate	District	Communities	Population
Duhok:	Akre	32	30,313
	Amedi	95	91,705
	Duhok	37	20,276
	Semel	12	8,650
	Shekhan	13	24,023
	Zakho	56	41,098
	Total of Duhok	245	216,065
Erbil:	Choman	73	26,654
	Dashti Hawler	12	2,105
	Erbil	1	3,000
	Khabat	0	-
	Koysinjaq	29	96,348
	Makhmur	0	-
	Mergasur	44	15,340
	Shaqlawa	31	69,971
	Soran	109	52,609
	Total of Erbil	299	266,027
Suleimaniya:	Chamchamal	44	53,138
	Darbandikhan	54	11,078
	Dukan	41	24,764
	Halabja	55	19,317
	Kalar	20	15,935
	Kfri	13	32,750
	Penjwin	95	31,419
	Pshdar	58	25,994
	Ranya	21	12,268
	Sharazur	5	2,375
	Sharbazher	147	31,318
	Suleimaniya	29	6,203
	Total of Suleimaniya	582	266,559
Kurdistan Region	**Total**	**1,126**	**748,651**

Table 6.6 shows the distribution of affected districts, communities and populations for each of the three governorates. It underlines the unequal presence of mine-affected communities among the three governorates. At the low end, Duhok counts for 248 mine-affected communities representing 22% of the total number. In the other hand, the governorate of Suleimaniya comprises 582 such communities representing 51% of all mine-affected communities.

A total of twenty-seven districts of the three governorates are mine-affected. The village survey shows a combined population of approximately 748,651 people living in the mine-affected communities, while another 2,809,750 people live in a mine-free environment. With a total population of approximately 4,910,742 in the three Governorates of the region, approximately 21 persons in every 100 people are actually living in a mine-affected community.

3.7 Settlement Type and Population Size

Mines and ERW primarily affect village communities (as opposed to urban ones). As demonstrated in Table 6.7 below, 976 out of the 1,022 mine affected communities are categorized as villages. However, other settlement types, e.g. collective towns, district centres, and sub-district centres are also very important in this regards. They account for more than half of the affected population, despite the fact that numerically they represent only a low proportion (4.5%) of all mine-affected communities.

Table 6.7. Affected communities and populations, by settlement type

Settlement Type	Affected Communities	Population Living in Affected Communities	Mean Population
Urban	5	149,400	29,880
Suburban	51	261,073	5,332
Compact Village	775	265,370	342
Dispersed Rural	232	56,884	245
Seasonal Settlement	47	5,064	108
Abandoned	16	10,860	572
Total	1,126	748,651	665

The population size spans four magnitudes, with the smallest community reporting only five people. Some villages report a population of zero inhabitants, for various reasons as indicated in Table 6.8 below. Internal displacement is given as the main reason for depopulation of these villages. The largest mine-affected community amongst those surveyed is the sub-district centre of Koysinjaq (located in Koya) which in turn is part of the Erbil governorate, claiming a total population of 96,348 people.

Table 6.8. Villages showing zero population

Classification of Villages with Zero Population	Number of Villages Showing Zero Population
Internally displaced persons IDP	35
Summer / Winter Village	12
Security	5
Unknown	11
Total	**63**

The affected communities and their populations by type of settlements are also shown in Fig. 6.2 below. Again, the figure shows that suburban and compact village areas are the areas with the highest concentration of affected communities, and also shows the much skewed population size distribution. This is consistent with expectations of such distributions in general. Although not all the parameters of the distribution for all communities in the region are known, there is no indication that affected communities follow a typical population size pattern.

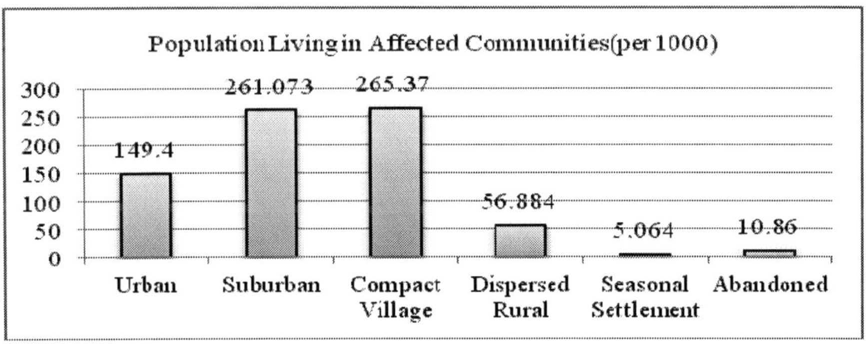

Fig. 6.2. Population size distributions by community-level

The collective towns (Fig. 6.2) comprise 3% of the total population consisting mainly of internally displaced people as a consequence of the Iran-Iraq war and the internal conflict between the Kurdish parties in two periods of time (1994–1995) and 1996, and have since then settled in collective towns. Although it may seem that the impact survey was biased in favour of small communities, this is not the case. All heavily populated communities were surveyed, but none of the very large cities (e.g. Duhok, Erbil or Suleimaniya city) were found to be affected by mines. This is due to the fact that mines were used mainly to block the progress of hostile forces in rural areas.

3.8 The Geographic Distribution of Affected Communities

The majority of mine-affected communities are in the southern governorate of Suleimaniya. However, there is a high concentration of such communities in the entire region, for example in the districts of Amedi (Duhok), Choman (Erbil), Sharbazher (Suleimaniya) and Penjwen (Suleimaniya). Figs. 6.3 to 6.6 below show the distribution of mine-affected communities by governorate and their classification by high, medium and low impact mine-affected communities in the region.

As shown in Map 6.1 and 6.2 below, clusters of mine-affected communities are clearly evident. They largely correspond to the history of armed conflicts in which landmines were used. For example, the communities that saw mines planted in the nineties during the Gulf War are for the most part in the Governorate of Duhok (i.e. in Fishkabor and in Sarsang). Mines used in the eighties during the Iran-Iraq war are concentrated along the north-eastern border of the country. Finally, the internal Kurdish political conflicts illustrate a random distribution of mined areas throughout all three governorates.

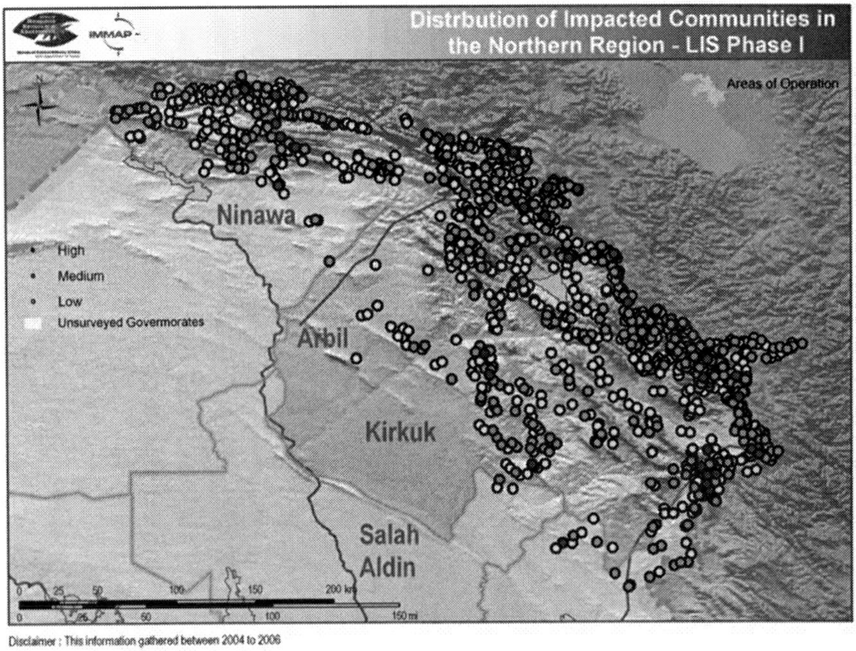

Map 6.1 Mine Affected Communities in the three governorates

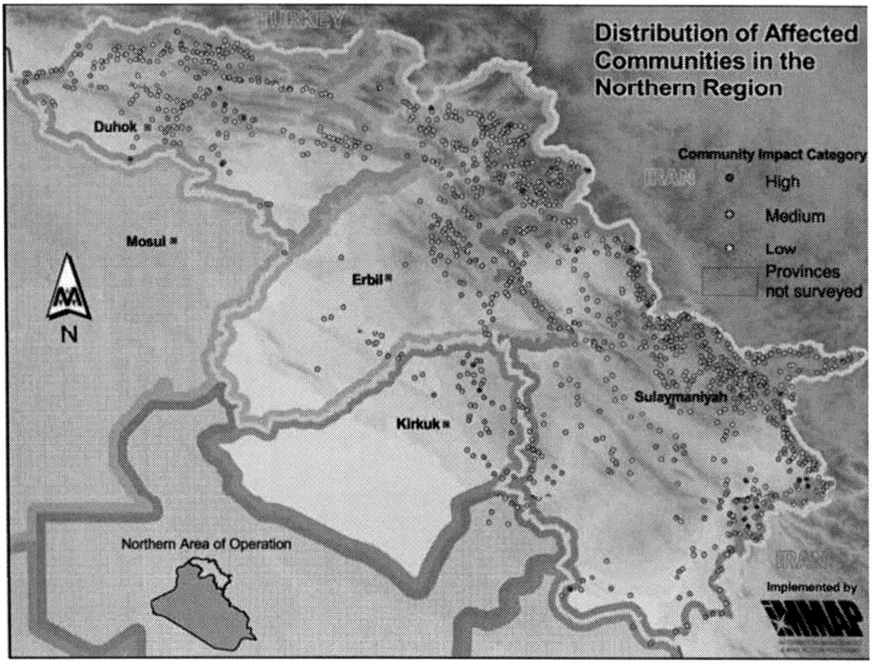

Map 6.2 Impact Communities in Iraq

3.9 The Mined Areas

The LIS have identified 3,026 distinct areas that are suspected to be mine and/or ERW contaminated areas. The data collectors, with the aid of key informants in the affected communities, have estimate that the suspected areas cover a total land surface of 776 square kilometres which is significant considering its relatively high intensive use by the rural population.

The size differences among those estimates are extreme. While the population range among the affected communities spans four magnitudes, seven magnitudes separate the smallest suspected area (8 square metres) from the largest (10.4 square kilometres). In fact, the differences are so large that a graphic representation of their absolute values is almost pointless. Instead, Fig. 6.3 groups the estimated suspected mine-contaminated areas by magnitude as captured during the impact survey. A very clear distribution appears in this case.

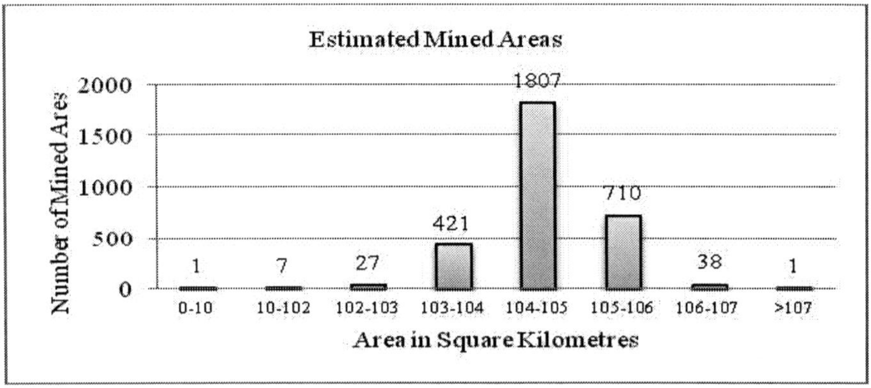

Fig. 6.3. The magnitudes of mined area surfaces

The mean size of minefields is 112,746 square metres (excluding the 236 minefields of unknown area). About 55.6% of the individual estimated suspected areas are between 10,000 and 100,000 square metres. A total of 23.1% are larger than 100,000 square metres and 14% of them are smaller than 10,000 square metres. Finally, the remaining 7.3% of the surveyed mined areas have unknown dimensions. These unknown dimensions are due to various reasons, such as being located within restricted areas, or that they are suspected mined areas that have been cleared by other agencies (e.g. Solidarity) or by villagers. Since some of those areas have been "cleared" by other agencies, including local organizations that do not adhere to the International Mine Action Standards (IMAS), they were in the process of being verified by the mine Action team.

Experience from the LIS indicates that a proportion of the surface estimates may be an overestimated. For instance, communities may use size to indicate the potential magnitude of their mine affected problems. They are meaningful in several respects, as a coarse indicator of the clearance challenge and also for the fact that these estimates are to a greater extent correlated with the probability of mine accidents. In such cases, only a technical survey will produce more realistic estimates of the contaminated surfaces.

As one would expect, the number of distinct mined areas in each community varies considerably, as illustrated in Table 6.9 below, this factor may be the outcome of two distinct processes: the actual laying of mines and the scattering of ERW, and/or the growing acquaintance, over time, of community members with the distribution of minefields. In other words, the 345 communities which claim one suspected area include an unknown number of situations where the community does not make the distinction, or did not care to point polluted areas out to the survey teams. It is with this caution that the following breakdown of communities by number of local suspected areas needs to be appreciated.

Table 6.9. Communities, by distinct suspected mined areas

Number of Communities	Distinct mined area in the Community	Percentage
417	1	37.0
273	2	24.2
185	3	16.4
91	4	8.1
55	5	4.9
40	6	3.6
16	7	1.4
12	8	1.1
14	9	1.2
6	10	0.5
17	More than 10 areas	1.5
Total: 1,126		**100.0**

As can be seen from Table 6.9 above, very few communities claim more than ten distinct areas, although this is relatively more common among higher impact communities. Fig. 6.4 below illustrates the distribution of high, medium, and low impact communities according to distinct mined areas. The high and medium impact communities show a much more even distribution than the low impact communities, which indicates that these communities have various numbers of mined areas around them. However, with low impact communities the distribution clearly shows that by far the most communities in this category have less than four mined areas.

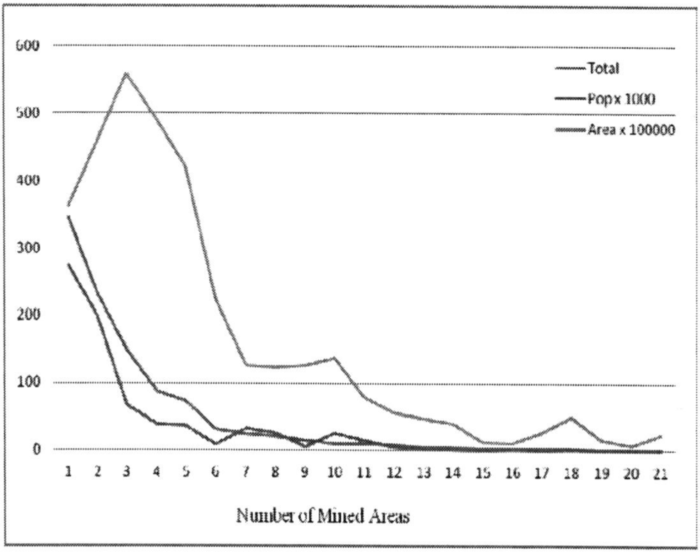

Fig. 6.4. Distinct mined areas by community (high, medium and low impact)

In Fig. 6.5 below, the number of mined areas is shown in relation to the communities, their population and the mined area. As expected, the curves of the communities and their population follow roughly the same path. However, when compared with the mined area, this relation becomes heavily skewed. From the graph it can be said that the bulk of the mined areas are located around communities with less than six distinct mined areas, and most of these communities are low impact communities (from Fig. 6.4 above).

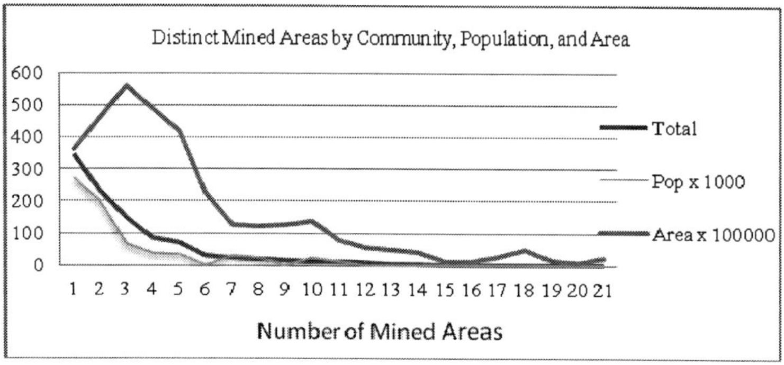

Fig. 6.5. Distinct mined areas by community, population and mined area

3.10 Mine Accidents' Victims

During the Impact Survey, 167 communities had a history of mine accidents harming 474 victims in the period of 1999–2000 and 2005–2006 (referred to as "recent incidents/accidents"). This data represents a comprehensive data set covering two different year periods. The following Table 6.10 summarizes the relevant victim figures.

Table 6.10. Victims of mine accidents

Period	Victims Killed	Victims Injured	Total Victims
Recent Victims (1999-2000)	108	207	315
Recent Victims (2005-2006)	60	99	159
Victims of Less Recent Date	1,033	1,937	2,970
All Victims	4,464	8,973	13,437

More information was sought on more recent victims, in part for reasons of interviewee recall. The figures on less recent victims are at best approximate, while the recent ones are believed to be more reliable.

3.11 The Severity of Impacts

The LIS has calculated a point score for each affected community expressing the severity of the various mine impacts. The score takes three major impact factors into account as follows:

- The type of munitions presents (landmines and ERW).
- The livelihood and institutional areas to which mines block access.
- The number of recent victims (from incidents of last two years).

The score ranges that qualify a mine affected community as being low, medium, or high impact has been calculated based on the thresholds shown in Table 6.11 below. These thresholds were established by the Survey Working Group (SWG),[11] and have been frequently applied in landmines impact surveys conducted in other countries.

Table 6.11. Score range for impact classification

Score range	Impact Classification
1–5	Low
6–10	Medium
11 and higher	High

The score is then used to classify the impact of landmines on communities as low, medium, or high. The method is explained in greater detail in Appendix A, titled "Process to Determine the Level of Impact (high, medium, and low)."

The actual scores range from 1 to 29 points. By definition, a score of 1 stands for a community that reported only the presence of some ERW without any blockages or recent victims caused by it, and the survey found only one such community. On the other extreme, Mami Khalan, in the sub district of Choman in the Governorate of Erbil, attained a score of 29. This community has nineteen minefields within its geographical boundaries, and had thirteen recent victims. Maps 6.2 to 6.4 below provide a list of mine-affected communities based on their impact.

Table 6.12. Frequency of impact scores

Impact Score	Total affected communities
1	1
2	10
3	92
4	270
5	287
6	216
7	99
8	75
9	30
10	11
11	5
12	9
13	5
14	1
15 and more	15
Total	**1,126**

It is observed from Table 6.12 on the frequency of impact scores, and from Fig. 6.5 that there are three communities recording scores of 25 and higher. On the other hand, nearly all of the 103 communities with impact scores from 1 to 3 have no socio-economic blockages.

All other mine-affected communities were given scores between these extreme values. However, as Fig. 6.6 indicates, most communities (96.6%) had a score between 2 and 10.

Fig. 6.6 indicates that the shape of distribution is heavily skewed to the low values, indicating mine-affected communities as having the presence of mines with minimal blockages. This characteristic, combined with education and marking and cleaning programs, has effectively reduced the affected communities' economic blockage and impact scores.

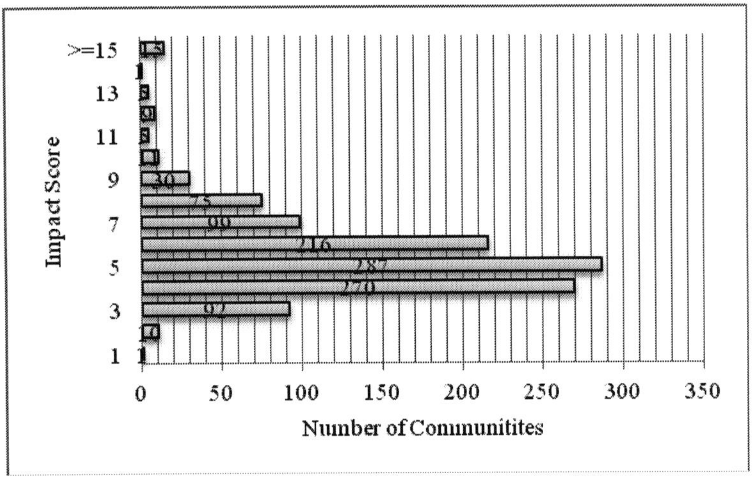

Fig. 6.6. The distribution of impact scores

A total of 788 or 77.1%, of the 1,022 mine-affected communities are classified as low impact communities. Another 199 mine-affected communities or 19.5%, are considered as medium impact, while the remaining 35 mine-affected communities, or 3.4%, are classified as high impact. The following graph (Fig. 6.7) illustrates the actual distribution of communities classified as low, medium, and high-impact communities.

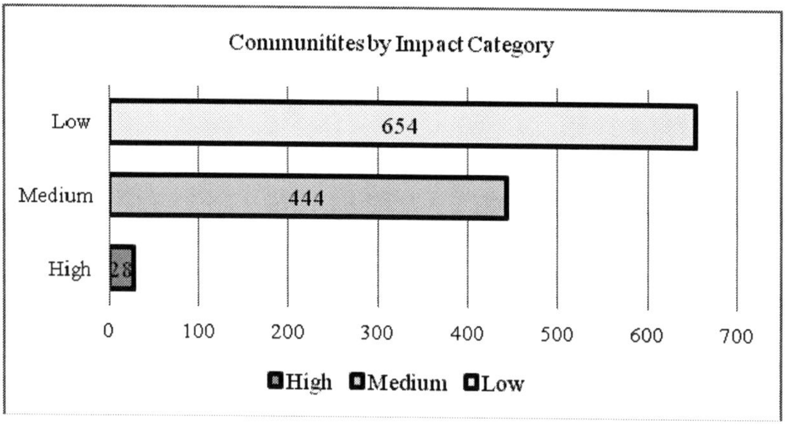

Fig. 6.7. Impact classification

3.12 Population by Impact Category

The estimated size of the population living in mine-affected communities is 748,651. As a result of the scoring and classification process, about 81,713 (11%) people are classified as living in high-impact communities. An estimated 241,026 (33%) live in medium impact communities and a remaining 425,912 (46%) of the population live in communities that the landmine impact survey rated as low impact.

Table 6.13 below summarizes the above-mentioned information. These figures deviate significantly from what is expected. Only 3.4% of the communities are classified as high impact communities but 21% of the mine-affected population lives in these communities. On the other extreme almost 77% of the communities are classified as low impact but only 46% of the mine-affected population live in these communities.

Looking at the individual thirty-five high impact communities it can be noted that it is indeed true that some of the larger villages fall into this group. Five of the thirty-five high impact communities have a population of greater than 10,000 with a total of over 135,000 people.

Table 6.13. Communities and population classified by impact categories

Impact Category	Communities	Affected Population
High	28	81,713
Medium	444	241,026
Low	654	425,912
Total	1,126	748,651

3.13 Landmine Victims

Demography of Recent Victims

The distribution of the 474 recent victims by age and sex (Table 6.14 below) indicates the groups of concern—for example, as many as 94.9% of the victims are male. Men ranging in age from fifteen to twenty-nine account for 45% of all victims. The number of victims between ages of five to fourteen account for 22% of all victims, which indicates that a large number of young children fall victim of mines and ERW. The number of female victims also falls predominantly into the age groups five to fourteen probably due to the same reasons as for the male children.

Table 6.14. Recent victims classified by age and gender

Age range	Male	Female	Total
0–4	4	1	5
5–14	95	9	104
15–29	214	4	218
30–44	71	1	72
45–59	43	3	46
60 and above	23	6	29
Total	**450**	**24**	**474**

The age distribution for male and female victims is almost identical, but the fact that men make up 94.9% of all victims can be explained by the imposed restrictive public movements of young women, especially in the rural areas. This difference is attributed to the local gender biased tradition.

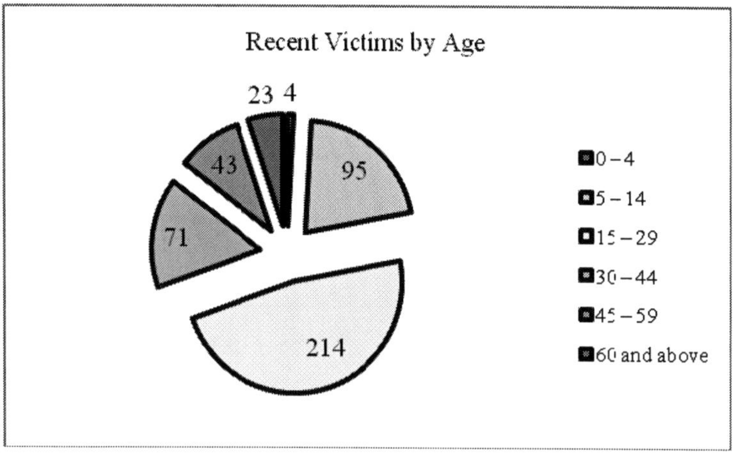

Fig. 6.8. Distribution of recent (2005–2006) victims by age

Given that young people are primarily responsible for farming and other agricultural activities such as husbandry, it is not surprising that they account for the majority of recent victims. Fig. 6.8 provides a breakdown of recent victims (in the last two years), combining both male and females. Again as expected, children, young cohorts and the working age population account for the largest shares of the total victims.

Moreover, Table 6.15 below offers a breakdown of recent victims by sex and occupation prior to the accident. It illustrates that the most frequent occupations having mine accidents are: children, shepherds, military, farmers, and workers, together account for 87% of all recent victims. This is consistent with the above findings shown by the analysis of recent victims by their age and gender.

When comparing the occupations of victims across males and females, of the 450 male victims, 67% are farmers, workers, children or shepherds. On the other hand, of the 24 female victims, only 1% are classified as children or infants, and 14 (3%) are housewives. Table 6.15 above indicates the frequent prevalence of domestic occupations among women. The impact survey does not provide more detailed information on those classified as "villagers."

Table 6.15. Occupation of landmine victims prior to accident, 2005–2006

Occupation before the mine accident	Male	Female	Total
Child/Infant	57	5	62
Shepherd	156	0	156
Military Occupation	50	0	50
Farmer/Peasant	66	1	67
Handyman/Worker	37	0	37
House wife/House keeper	0	14	14
Official employee	20	0	20
Unemployed	7	0	7
Driver	7	0	7
Villager	2	0	2
Education/Training	28	2	30
Trading	13	0	13
Others	7	2	9
Total	**450**	**24**	**474**

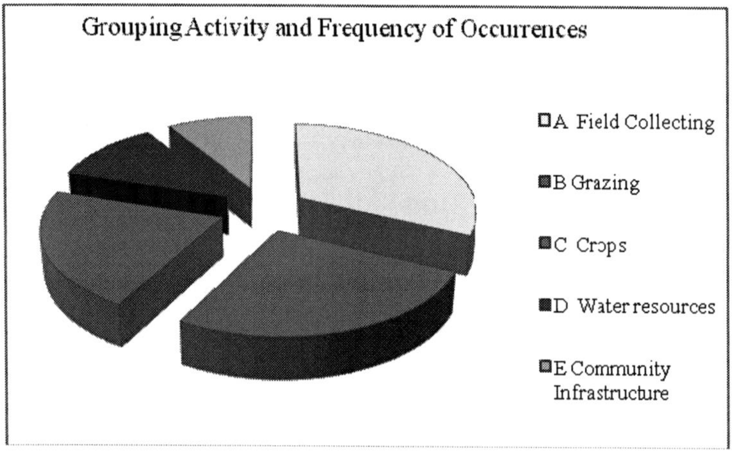

Fig. 6.9. Recent victims by community

Fig. 6.9 illustrates the distribution of the 474 recent incidents, encompassing 273 mine-affected communities. The numbers provided a basis for comparison of landmine fatalities across communities. This is important considering resources allocation by the degree of contamination and in proportion to the actual accident rate. Such policy enhances their inclusiveness, trust, participation in development processes and general well-being on affair allocation basis.

Accidents and the Consequences

Similar to the occupational breakdown, herding and other agricultural related activities are strongly present among the activities performed at the time of the accident (see Table 6.16 below). In addition, nearly 30.8% of the victims were involved in, or close to somebody, tampering with mines. This is consistent with the data gathered from communities where mine clearance activities were conducted. It should be noted that tampering with munitions is the most prominent cause of accidents for males and the second most prominent for females. Perhaps this can be attributed to "mine hunting," whereby villagers extract metal and explosives from landmines and UXOs found lying around for recycling purposes.

Table 6.16. Activity at time of accident

Activity at time of accident	Male	Female	Total
Tampering with/destroying mine/UXO	101	3	104
Walking, running, sitting, picnic, driving, travelling, riding, nothing, normal activity	90	9	99
Playing	18	1	19
Keeping, grazing animals, staking livestock	139	0	139
Digging in the ground to catch animals, infrastructure purposes, agricultural purposes, farming	28	2	30
Military	9	0	9
Making a fire for cooking, clearing land, destroying garbage	2	2	4
Collecting wood, thatch, straw, fruits, nuts, food, metal, herbal medicines and cutting vegetation	41	20	61
Other	9		9
Total	**437**	**37**	**474**

Out of the 474 recent victims shown in Table 6.17 below, 159 died, of which 151 were male. The fatality rate of each accident for females is 33.4% while for males it is 33.6%. The difference is statistically insignificant as the impact of munitions is not different, but the frequency by gender differs, linked to their occupation and nature of activities.

Table 6.17. Mine accident fatalities by gender

Victims and accident outcomes	Male	Female	Total
Killed	151	8	159
Injured	299	16	315
Total	**450**	**24**	**474**
Fatality rate	33.6%	33.4%	34.0%

In addition to that, Table 6.18 below indicates how mine victims died. Over 93% were killed instantly, regardless of gender, while a minority (7%) passed away later as a result of mine accidents that caused injuries and subsequent complications mainly attributed to infections.

Table 6.18. Percentage victim deaths by gender

Victims and accident outcomes	Male	Female	Total
Died later	11	0	4
Killed instantly	140	8	148
Total	**151**	**8**	**159**
Percent killed instantly	92.7%	100%	93.0%

Moreover, the occupational and health status of the survivors during the post-accident period is shown below in Table 6.19. A large number of victims become unemployed or disabled due to trauma and injuries reducing their working capacity. Some other categories such as shepherds, villagers, teachers and government officials, return to labour market and their jobs after a rehabilitation period.

Table 6.19 indicates that nearly 20% of the survivors are handicapped and 18% remain unemployed. 3.5% are children, and are most likely to have a disability or a long-term negative effect on their parents and own future levels of unemployment. However, if the children are removed from the statistics, in order to account for the labour force, 19% of the survivors remain jobless, 20% are disabled and only 60% can continue with normal jobs. The underdeveloped infrastructure for the handicapped is another factor reducing the low mobility and labour market participation of the disabled on an equal basis as of those non-disabled. According to the LIS report, there are no special programs providing employment opportunities to the disabled in the public sector.

Table 6.19. Mine accident survivors, by sex and occupation

Occupation of survivors	Male	Female	Total
Child	8	3	11
Unemployed	55	1	56
Shepherd	43	0	43
Handyman/Worker	17	0	17
Farmer/Peasant	22	0	22
House wife/House keeper	0	5	5
Official Employee	21	0	21
Villager	31	1	32
Teacher	18	2	20
Retired	2	0	2
Shop keeper	3	0	3
Unable to work (Disabled)	58	3	61
Other	13	0	13
Total	**291**	**15**	**306**

4. Analysis of Blocking Components

4.1 Communities Claiming Certain Types of Blockages

The Key informants in the affected communities, such as Anjuman, Mukhtar, and local landowners have identified four major areas to which mines and ERW most frequently block access as follows:

- Grazing.
- Crops, firewood, hunting, and wild plants.
- Drinking, irrigation, shelter material, and road access.
- Power, bathing, fishing, and town point features.

Table 6.20 below shows the frequency and percentage of communities that reported mine presence as some form of blockages in these specific areas. This table presents a summary of blockage factors grouped by major impact type.

Table 6.20. Communities affected by mine blockages

Mines Block Access to	Frequency of occurrences (Communities)	Percentage (%)
Grazing	316	26.2
Crops	262	21.7
Firewood	142	11.8
Wild Plants	128	10.6
Hunting	91	7.5
Irrigation	82	6.8
Drinking	43	3.6
Power	36	3.0
Road Access	31	2.6
Shelter materials	23	1.9
Housing	20	1.7
Fishing	7	0.6
Recreation	7	0.6
Religious Site	7	0.6
Bathing	5	0.4
School	2	0.2
Factories	1	0.1
Medical	1	0.1
Market	0	0.0
Bridge	0	0.0
Total blockage	**1,204**	**100.0**

It should be mentioned that only 450 communities experience blockages. However, the total adds up to 1,204 community cases and is not limited to 450. This is due to the fact that multiple forms of blockage instances can be recorded for one community.

The landmine impact survey refined those categories of the key informants by grouping activities according to their similarity and the importance of the blockage factor. For example, the two activities that are mainly blocked (as can be seen in Table 6.21 below) are grazing animals and growing crops. These two activities should therefore form separate categories. Three more categories were created according to these criteria, and the new grouping process is illustrated in Table 6.22 below.

Table 6.21. Categorizing blockage factors

Mines Block Access to:	Grouped into new Categories	Frequency of Occurrences (%)
Grazing	Grazing	26.2
Crops	Crops	21.7
Drinking	Water resources	3.6
Irrigation	Water resources	6.8
Bathing	Water resources	0.4
Fishing	Water resources	0.6
Firewood	Field Collecting	11.8
Hunting	Field Collecting	7.5
Wild Plants	Field Collecting	10.6
Shelter materials	Field Collecting	1.9
Housing	Community Infrastructure	1.7
Road Access	Community Infrastructure	2.6
Power	Community Infrastructure	3.0
Recreation	Community Infrastructure	0.6
Factories	Community Infrastructure	0.1
Medical		0.1
Market		0.0
Religious Site		0.6
School		0.2
Bridge		0.0
Total Frequency of Occurrence		**100.0**

The frequency of occurrence shown in Table 6.21 above in disaggregated form grouped by blockage factors into more aggregated blockage factors is reported in Table 6.22 below. The agriculturally related factors including field collecting, grazing and crops are among the dominating blockage factors.

Table 6.22. Frequency of occurrence grouped by blockage factors

Group	Grouped Blockage Factors	Percentage
A	Field Collecting	31.8
B	Grazing	26.2
C	Crops	21.7
D	Water resources	11.4
E	Community Infrastructure	8.9
Total		**100.0**

A regrouping of the activities and frequency of their occurrence is also shown in Fig. 6.9 by their shares. As mentioned above, the figure shows a clear picture of the dominance of grazing, crops, and field collecting activities accounting for the largest share of landmines accidents in the region.

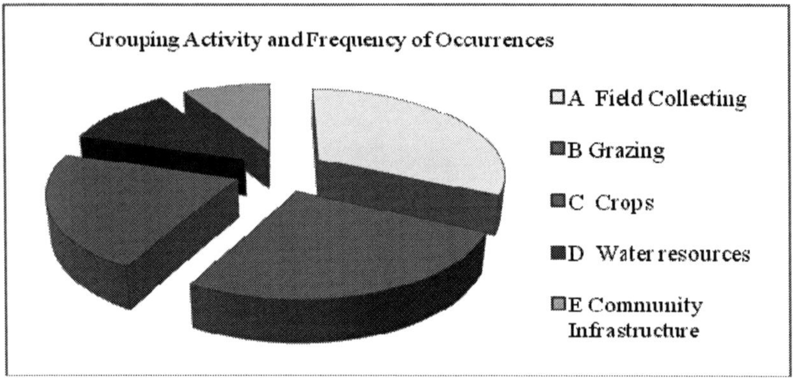

Fig. 6.9. Frequency of occurrence by blockage factors

Based on the information presented above regarding frequency and distribution of accidents by individual characteristics, several findings stand out as follows:

- The category most affected by landmines is field collecting. This is consistent with the fact that the rural community is very dependent on their natural surroundings to provide them with daily necessities like energy and water.
- As a single activity, grazing is the category most affected by landmines, particularly around strategic hillsides and mountaintops that were former military positions. Given the nomadic nature of the shepherds and their prominence throughout the region, this figure suggests that one out of every three villages is affected by landmines impeding access to grazing land.
- Another single activity affected by mines is growing crops, which is again consistent with the social and economic activities of the majority of the rural population. Field Collecting, grazing, and crops make up the bulk of the activities that are affected by landmines in the Kurdistan region and clearly indicate the rural nature of this problem.

- Water resources and community infrastructure mostly relate to the urban population of larger villages, sub-districts, and district centres. There is a steep descent in the number of mine-affected communities from the previous categories, which is again consistent with the rural nature of the problem. Institutional features such as schools, religious sites, medical and recreation sites in addition to bathing and fishing sites are represented in these categories. This may not be unusual for mine-affected communities that are predominantly rural, and which either have few such institutions or have enough space to secure access to them around suspected areas.

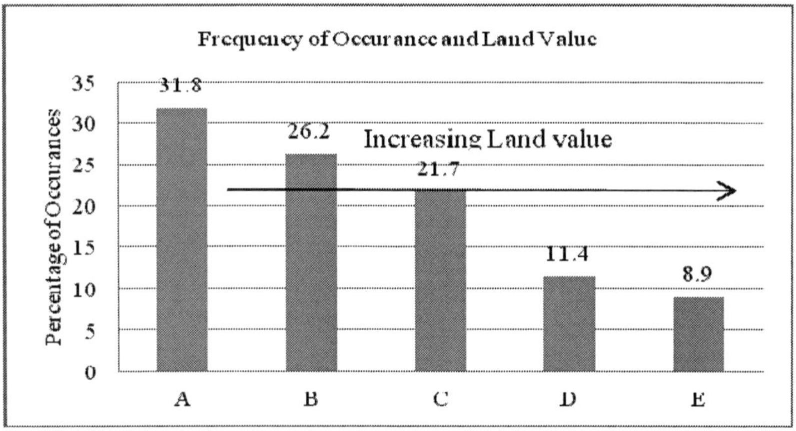

Fig. 6.10. Frequencies of occurrence and land value

There are five major prevalence levels as shown in Fig. 6.10 and corresponding subjective land value can portray the blockage factors:

- 31.8% of the occurrences grouped as “A” within mine-affected communities reduce access to low value land holdings predominantly used for field collecting.
- 47.9% of the occurrences grouped as “B” and "C" within mine-affected communities reduce access to intermediate value land holdings.
- The remaining 20.3% of the occurrences grouped as “D" and "E” within mine-affected communities reduces access to high value land.

Table 6.23. Populations, quantity of mined areas and estimated surface area associated with blockage type

Type of Impact	Communities Affected	Population	No of Mined Areas	Estimated Area in sq. kilometres
Crop	262	152,920	553	62,138,914
Grazing	316	199,585	733	102,035,754
Drinking	43	41,940	61	10,470,063
Irrigation	82	71,185	137	19,468,636
Bathing	5	26,130	7	1,340,600
Fishing	7	26,420	13	1,646,600
Firewood	142	125,285	268	46,180,114
Hunting	91	74,450	165	30,934,045
Wild Plants	128	91,115	238	43,960,775
Shelter	23	38,595	26	3,084,019
Housing	20	12,970	26	5,810,829
Road Access	31	18,520	48	5,272,312
Power	36	24,760	45	2,253,383
Recreation	7	25,895	9	465,300
Factories	1	75	2	254,000
Medical	1	35	1	16,000
Market	0	0	0	0
Religious Site	7	7,340	8	721,500
School	2	985	2	16,200
Bridge	0	0	0	0

It is worth mentioning that the determination of land value is not empirical, but is based on subjective and intuitive observations by local experts. These figures indicate that the impact of landmines on the population is mainly focused on the people living in areas with a low land value, i.e. rural areas. It also implies that the landmine problem affects the lower income group of the population, who are dependent on their natural surrounding areas to form a life.

The size of the population affected by a particular problem as well as the numbers and surface of the suspected mined areas involved can be determined, as indicated in Table 6.23. It should be kept in mind that a given mined area may be blocking several resource types, and thus the factor of overlap between these categories is considerable. This data is consistent with the observations presented above.

4.2 The High, Medium, and Low Impact Distribution

Table 6.23 can be refined further to form Table 6.24 below to indicate the distribution of blockage factors over communities classified as high, medium, and low impact communities. The table is grouped into the same categories, A through E, as indicated by Tables 6.22 and 6.21 above.

It should be noted that individual mined areas in Table 6.24 do not have a high, medium or low impact, but that the existence of the mined areas block certain basic needs to a neighbouring community. Together, the number of blockages, the number of incidents, and the existence of ERW, AP, and AV mines contribute towards estimation of an impact score for the community. The number of mined areas therefore may or may not have an influence on the impact score, depending on how many factors are blocked by these minefields. Therefore, it is misleading to interpret Table 6.24 as categorizing mined areas into impact categories.

As explained before, the number appearing in the sub column "H" in the column "No. of mined areas", merely refers to the number of mined areas that block access to high impact communities (in categories A,B,C,D and E), it does not attempt to classify mined areas into impact categories.

Table 6.24. High, Medium and Low Impact Distribution of Blockage Factors

Category	Communities Affected by blockage factors				Population affected by blockage factors			
	High	Medium	Low	Total	High	Medium	Low	Total
A	39	172	173	384	46,910	216,970	65,565	329,445
B	22	152	142	316	17,560	146,800	35,225	199,585
C	24	151	87	262	31,295	81,545	40,080	152,920
D	26	93	18	137	29,650	120,635	15,390	165,675
E	21	50	34	105	15,240	47,970	27,370	90,580
Total	132	618	454	1204	140,655	613,920	183,630	938,205
True Count	28	444	654	1126	81,713	241,026	425,912	748,651
Ratio	3.77	3.11	0.58	1.18	0.91	2.53	0.54	1.27

Category: A- Field collecting, B– Grazing, C- Crops, D- Water resources, E– Community Infrastructure

It is noteworthy that, as expected, on average more blockage factors occur in the communities rated as high and medium, with the majority of occurrences in the medium impact group. This can be explained by making use of ratios. Looking at the ratio row in Table 6.24 it can be seen that on average, the 28 high impact communities have 3.77 blockages per community, the 444 medium impact communities 3.11, and the 654 low impact communities have 0.58 blockages per community. These numbers indicate that the scoring method based on impact and blockages has the desired property and that it addresses the negative impact of minefields on communities and not the simple number of mined areas.

In analyzing the ratios of affected population to the true population, a ratio of 1:1 would imply that, on average, every person is affected at least once by a blockage factor. The ratios are therefore also reflections of the impact that the blockages have on the population. The most prominent blockage impact is on the population of the medium impact group with a ratio of 2.53. The population number "affected by blockage factors" should therefore not be confused with the population living in medium impact communities, because one person can be affected by multiple blockage factors from one mined area. The figures for population and communities in Table 6.24 above therefore reflect the impact of blockage factors and not only the exact number of communities and households.

Analysis of the number of minefields that contribute towards blockage numbers yields a total of 2,342 minefields as having blockage impacts. This number also includes multiple occurrences, because one minefield can block several factors. Of the 3,026 surveyed minefields in 1,126 impacted communities, only 1,926 minefields in 450 mine-affected communities contribute towards blockage factors. This results in 1,322 minefields in 572 mine-affected communities that do not block any activity.

Inspection of the ratios for the number of mined areas that contribute to the blockage factors show that with 1.35 and 1.34 in high and medium categories respectively, minefields in these categories block more than one factor on average. In the low impact category, the ratio is only 0.34. This explains the 1,322 minefields that do not block any activities. This analysis again shows that the method of scoring based on impact and blockage has the desired property and explanatory power.

5. Casualty Analysis

The survey action centre which has been responsible for a majority of the landmine impact surveys that are conducted worldwide, has assisted with the analysis of the LIS data, to produce the results contained in this section.

Other more advanced methods will be used in forthcoming chapters of this book, where we study distribution of the data and probability of accident and outcome of accidents. Through regression analysis, we attempt to estimate the marginal effects of different causal factors.

A form of analysis commonly used in landmine impact surveys in other countries was applied to the Kurdistan region that being the regression of the number of recent victims on suspected causal factors. The methodology is presented in Appendix A.

From Map 6.1 above, we can see that multiple clusters of communities with some recent victims (red dots; larger dots representing more victims) of the Kurdistan region are many, but not all of the clusters are near the Iranian border. Statistical analysis reveals that the distance to the nearest other community with some recent victims is a better predictor of victim numbers than the distance to the border.

The main findings of the regression models include:

- The population size of the affected communities is the strongest determinant both for the chance to have no victims at all, and to have more or fewer for those who do have victims. Larger communities manage less often to sustain no accidents, and they tend to have more victims than smaller communities. This population effect seems to be stronger than in other countries, perhaps justifying efforts as being more strongly focused on larger communities.
- While at first glance of a victim map, communities close to international borders may appear to have more victims, and this distance is not important for the group of communities that underwent a formal impact survey. For the number of victims, the distance to the nearest other community with some recent victims is strongly discriminating. This is a clustering factor that was found to be very strong as well in other country surveys. The results suggest that the next phase of mine action resources should not be allocated to individual high-impact communities, but rather to clusters of communities.
- Relative risk is an effective and simple measure of prevalence of risk in a community. The number of victims per 100,000 person over years, a common rate definition used in other surveys, is 15.1, keeping in mind that recent victims include all within a two-year period and assuming that the average family size in the Kurdistan region is 7.0. This figure is calculated as (474 recent victims* 100,000) / (2 years*147,414 families*7.0 persons per family).The

predicted number of victims increased (as expected) with the population in the community, but the risk standardized to person per years decreases sharply. This latter result is not surprising, as the possible cause may be due to private property. Thus, the relative risk for the individual decreases with population size. The results are also seen in other countries, and it is nevertheless important policy wise:

- Recommendations may vary depending on whether one looks at the absolute numbers of victims, or at the risk.
- For several reasons—fear of discriminating against small, already neglected communities being one of them—the impact score metric does not take size of population into account.

- Also worth noting is the high proportion of affected communities reporting no resource blockages. This is much more frequent in the Kurdistan region than in other surveyed countries. The sheer number of blockage types checked, however, has some influence on victims, but it is not clear to what extent.

6. Multiple Blockages Types

6.1 Blockage Types by Frequency

The area of blocked access, number of communities affected and their percentages are shown in Table 6.25 below. The different areas are further disaggregated into their sub-areas and their magnitude of affect and shares reported. Again, pasture, crop land and other land uses as well as water usage areas dominate the sample.

Note that the frequency of the most common blockage type is access to grazing land at 98%. Grazing land often coincides with alternative use in form of its utilization for military strategic use.

Table 6.25. Communities with access blocked

Area of Blocked Access	Communities Affected	Percentage
Any Kind of Pasture	1,103	98
Migratory Pasture	748	66.4
Fixed Pasture	621	55.2
Crop Land	609	54.1
Rain-fed	467	41.5
Irrigated Crop	229	20.3
Non-Agricultural Uses of Land	888	78.9
Wild Food	798	71
Firewood/Charcoal	551	49
Wild Herbs	79	7
Building Materials	22	2
Other	254	23
Any Water Usage	144	12.8
Drinking Water	94	8.3
Other Water than Drinking	141	12.5
Any type of road	131	11.6
Roads to Administrative Centres	6	0.5
Other Roads	128	11.4
Infrastructure	20	1.8
Housing	9	0.8
Other Infrastructure	11	1
Total Communities	1,126	100

6.2 Communities with No or Several Blockages

It is not surprising that more than half of all surveyed communities reported that their mine and ERW contamination did not cause any of the suggested resource types to be blocked. The number of blockage types checked is reported in Table 6.26 below.

Table 6.26. Blockage type checked

No	Frequency	Percentage	Cumulative Percentage
0	571	55.9	55.9
1	148	14.5	70.4
2	159	15.6	86.0
3	74	7.2	93.2
4	40	3.9	97.2
5	14	1.4	98.5
6	8	0.8	99.3
7	4	0.4	99.7
8	2	0.2	99.9
9	1	0.1	100.0
Total	1021	100.0	

The high portion of zero-resource blockage communities in the Kurdistan region is much higher relatively than in other countries. It may be the most surprising finding of the current landmine survey.

6.3 Number of Blockage Types and Recent Victims

Table 6.27 below shows the blockage type and recent victims. There appears to be a significant influence of the number of different blockage types checked by communities and their mean numbers of recent victims. That is, communities suffering heavily from more types of blockages have a greater chance of having recent victims (see Table 6.27).

This should qualify the sheer number of blockage types for consideration as a policy relevant variable. An important issue in this regards is how to utilize the information, which is not immediately obvious.

Table 6.27. Blockage type and number of recent victims

Number of Blockage Types Checked	Number of recent victims	Mean number of recent victims
0	571	0.25
1	148	0.32
2	159	0.27
3	74	0.35
4	40	0.75
5	14	0.57
6	8	0.63
7	4	1.00
8	2	2.00
9	1	0.00
Total	**1021**	**0.31**

7. Summary of Past Mine Action

This section summarizes past mine action activities in areas of survey and clearance operations, as well as victim assistance and mine awareness based on other data sources external to the impact survey (clearance, victim assistance and mine risk education).

7.1 Previous Mine Action in Areas of Survey and Clearance Operations

This section provides a summary of the mine advisory group activity. The Kurdistan region has suffered protracted conflict for more than twenty-five years, with the most recent in 2003 leading to further landmine contamination and the extensive delivery of munitions, including cluster bombs, as well as huge numbers of abandoned ammunition stockpiles left behind by the surrendering Iraqi army.

The mine advisory group started operations at the beginning of 1992, its headquarters based in Erbil with operations bases in Dahuk, Diyana, Chamchamal and Suleimaniya. The program had nineteen clearance teams, six conventional weapon destruction emergency response teams, two mechanical teams, one mine detection dog team, and fourteen Community Liaison teams. Clearance statistics (July 2008 to June 2009) show evidence that thirty minefields and fifteen battle area clearance sites were cleared amounting to 1,350,167 square metres of land. Additionally, thirty-six new minefields were identified and demarcated, totalling

2,153,214 square metres. The released lands have been used for agriculture and the construction of housing blocks, roads, and bridges.

Moreover, the knock-on effect of cleared land facilitated infrastructure and development projects including the maintenance of irrigation channels.

Within the three governorates of the Kurdistan region, six types of groupings have been involved with the removal of landmines—Mines Advisory Group (MAG), Norwegian Peoples Aid (NPA), a local NGO known as Solidarity, local villagers, UNOPS-MAP and others (cleared by Iraq, Iran or allied countries). The Mines Coordination Offices (MCOs) of the Local Authorities, established with the support of the MAP, serve as the focal point for coordination of activities.

MAG began to identify suspected mined areas in 1992, and conducted an extensive permanent marking program in 1996–97, still evident today. As of October 2002, MAG has identified and surveyed approximately two-thirds of the minefields in the three governorates. It continues to be actively involved in demining and mine risk education. NPA established EOD teams in 1995, and has continued with EOD and more recently demining, particularly in the Mawat sub-district of the Suleimaniya governorate. NPA has cleared and handed over to the local population a total of 440,000 square metres, while disposing of some 3,100 landmines and 64,500 pieces of UXOs.

Those cleared minefields by Solidarity, villagers and others are considered as suspected areas mainly as a result of the lack of adherence to international standards. This represents 472 suspected mined areas that are presently being addressed with additional resources to verify their clearance of which 236 have no recorded area.

Since the UNOPS-MAP began operations, it has cleared 91 minefields, of which 81 have been assessed during Post Clearance Impact Assessments. Table 6.28 below shows the summary of past and current activities.

Table 6.28. Summary of Past and Current Activities[12]

Organi -zation	General Survey	Impact Survey	Verific -ation	Technical Survey	Clearance in progress	Clearance Complete	PCIA
MAG	2,233		180		20	180	
NPA	62		18		5	18	
Solidarity	unknown		66		unknown	66	
Villagers	unknown		386		unknown	386	
Others	unknown		20		unknown	20	
UNOPS	3,444	3,248	0	109	112	91	81
TOTAL	3,444	3,248	670	109	137	761	81

Notes: The number of clearances in progress for UNOPS includes temporarily suspended minefields that are exclusively winter tasks. UNOPS implements impact surveys, technical surveys and post clearance impact surveys whereas other organizations are believed not to have these processes.

7.2 Previous Mine Action Activities in the Areas of Victim Assistance

Data collection on victims during the general survey was developed prior to the existence of a comprehensive victim assistance component within the frame of mine action program. This data captured incidence of mines/ERW accidents, basic data on victims and information on action taken immediately after the accident. No information was collected on the exposure of the victims to services available except where the victim was taken from the site of the accident. Most of the information gathered on victims relates to the planning and evaluation of mine awareness activities.

Notwithstanding these factors, general aggregate data available is useful in drawing profiles of known victims, as well as for projecting future services. Victim assistance relied mostly on data available from the consolidated database created for the host of service centres presently serving these victims. However, other profiles were compiled for planning purposes, by age, gender, and urban/rural locations from the impact survey data. The condition for the most vulnerable group, children, is described below:

- *By age:*
  - Out of 13,437 victims registered, 5,730 of them (43%) were subjected to accidents before reaching the age of 18.

    - Among these 5,730 cases, 9% were five or under, 36% were between six and eleven, and 55% were between twelve and eighteen when the incident happened.
- *By gender:*
    - Among the 5,730 child victims, 90% were male, and 10% were female.
    - Within each gender group, 42% of the male victims subjected to accident before reaching the age of eighteen, while 50% of the female victims had their accident before eighteen.
- *Children as indirect victims:*
    - 40% of the victims were children. Among the adult victims, 77% were parents of more than three children.[13]

The other significant observation is that for clearance purposes, it is assumed that densely populated urban centres do not face a significant mine threat and consideration was given only to stockpiles of UXOs in these areas. However, the numbers of UXO victims from these areas cannot be ignored and data on actual recent cases demonstrate that a sporadic distribution of UXOs in these areas poses a serious threat, especially to children, and it is responsible for approximately 10% of all ERW victims per month.

The recently completed network of MAP victim assistance services in the three governorates provides a complete range of services in accordance with international guidelines for victim assistance, such as emergency services, prosthesis, and vocational training. Production and service provision rates demonstrate that most victims are being served according to their needs, there are no unreasonable waiting lists at the centres, and that approximately 40% of old victims need continuous additional services after re-integrating into their communities.

It should also be noted that the victim assistance services are providing benefit to the population as a whole. In the year 2000, among the new admissions of patients under nineteen registered by the two centres:

- 4% were suffering from Mine injuries;
- 12% were suffering from UXO injuries;
- 19% were suffering from bullet injuries, among which 78% were aged fifteen to nineteen.
- 65% were suffering from burn injuries (caused by hot water, tea, kerosene, electricity). Almost all of these children were younger than fifteen.

7.3 Previous Mine Action Activities in the Areas of Mine Risk Education

During the landmine impact survey, limited information was collected regarding previous mine risk education activities. MAG executes an extensive mine risk education program since 1992, focusing particularly on school teachers. However, with the establishment of local mine risk education NGOs in 2001 by the MAP, UNOPS mine risk education Teams have visited 600 communities, covering approximately 45% of the mine-affected communities as well as neighbouring villages not directly mine-affected and approximately 125,000 people were beneficiaries to these visits as indicated in Table 6.29. It is too early to measure the impact of the interventions so no clear conclusions can be drawn at this time. In forthcoming chapters more detailed analysis of the updated data will be conducted based on more advanced statistical methods.

Table 6.29. Summary of Mine Risk Education

Mine Risk Education Performance	Beneficiaries
Number of communities visited	601
Number of male beneficiaries	28,322
Number of female beneficiaries	29,272
Number of child beneficiaries	66,575

8. Summary of the Landmine Impact Survey in the Kurdistan Region

This chapter provided an introduction and an overview of the mine action in the Kurdistan region. The focus is on the basic original approach and how it was changed to an integrated mine action approach, and the resulting differences in their outcomes. The background and implementation, outcomes and lessons learned from the integrated approach is further discussed. In addition we presented detailed summary statistics covering the mine action resources, local capacity building, fields and communities have been classified and victims categorized by their gender, age and professions.

The mine action in the Kurdistan region illustrates the problems related to mine contamination issues. In general, it focuses on transfer and use of mines, mine awareness education programs, mine-victim assistance, survey of mine-suspected areas, and mine-clearance activities. These activities incorporate all the basic elements of dealing with mine problems.

In addition, there are numerous supporting activities which are crucial for the success of mine action.

The basic task of the Kurdistan mine action centre is to plan and implement mine-clearance operations in the region. To fulfil the national tasks, numerous supporting tasks and human skill development and organization structure and capability enhancing activities have been performed. The program consists of three core components: clearance, mine risk education, and victim assistance. These core activities previously planned their work largely in isolation of each other. The aim of the integrated program in 2003 was to coordinate all of the three core activities in order to address all impact factors on each individual community.

In order to implement the integrated approach program, UNOPS teams were made up of mine clearance specialists, mine risk education specialists, and victim assistance specialists who worked together with villagers to address the reduction of the landmine impact. In this approach, an integrated plan for clearance, victim assistance, and mine risk education was developed in collaboration with local services, contractors, and the mine coordination office. The use of the Mine Action Information System (MAIS) supported the implementation plan through collection, management and supply of vital data. Moreover, the implementation process consisted of four stages including planning, tasking, execution and analysis of results. Introduction and implementation of the integrated approach has led to changes in working procedure and outcome, and many positive lessons learned.

The UNOPS Mine Action program (UNOPS-MAP) has revised its priorities of independent minefields to focus on a community priority and size of mine-affected areas, resulting in the transfer to a cluster tasking approach. As a result of these activities, through a combination of marking, clearance and MRE, the landmine victim incidence rate fell by over 50% between 1999 and the end of 2002. In this way, MAP focused on growth in its activities via the national capacity development program in order to address the landmine and ERW contamination problem and provide better long-term sustainability in the effective demining of the Kurdistan region.

The landmine impact survey (LIS) in the Kurdistan Region was conducted by UNOPS-MAP in late 2002. It was then followed by a LIS for Iraq in general, which was conducted over a two year period, ending in late 2006. The LIS confirmed and complemented the original general survey conducted within the region. The general survey was conducted during the years 1998 to 2000, building in turn on data collected during the previous years. The two surveys' comprehensive coverage have served as complementary to each other and have provided a detailed picture of the magnitude of the mine problem in the region. The two different periods of LIS were conducted to identify specific blockages to community's social and economic activities.

A number of 25 to 27 districts within the Kurdistan region are mine affected—97 to 124 sub-districts, and 1,126 of 4,281 communities (villages), and an estimated total of 748,651 persons live in the impacted communities with a total population of 3.5 million. The governorate of Suleimaniya contains the majority of mine-affected communities (582 out of 1,126) with a mine affected population of 266,559 people. Moreover, the LIS report identified a total of 3,024 distinct areas suspected of contamination, which were assessed with respect to different blockage factors. With a total population of approximately 4.9 million in the three governorates of the region, approximately 21% live in a mine-affected community.

In the design of the clustering system, focus was concentrated on a group of mine-affected villages. Village clusters were developed based on the classified mine-affected villages ranked as high, medium, and low. The three activities mainly denied land in the region were non-agriculture uses, pasture and agriculture areas, accounting for about 79%, 98%, and 54% share of the respective land area. Pasture areas are the most affected land category by landmines, particularly around strategic hillsides and mountaintops that were former military strategic positions. Given the nomadic nature of the shepherds, this result suggests one out of every three mine impacted communities is strongly affected by impeded grazing land areas.

Completion of the landmine impact survey resulted in a comprehensive set of mine related socio-economic impact data. This data allowed the mine action community to develop effective plans that target areas posing the greatest threat and to identify communities bearing the greatest impact. With a strategically focused effort, the impact of landmines can be reduced more rapidly and efficiently, providing the greatest positive impact on the living conditions in the contaminated communities.

The result of the LIS indicated that all three governorates of the region suffer from many adverse consequences of landmines and ERW contamination. Although these impacts are felt throughout the region, the existence of dense "clusters" of highly and moderately impacted villages suggested the need for a well-planned and targeted set of initiatives that could bring about considerable improvements in the situation in a short period of time. The majority of mine-affected communities are in the southern governorate of Suleimaniya. However, there is a high concentration of mine-affected communities in the entire region. The distribution of mine-affected communities by governorate and their impacts classification is reported.

The clusters of mine-affected communities are clearly evident. They correspond largely to the history of armed conflicts in which landmines were used. For example, communities that saw mines planted in the nineties during the Gulf War are for the most part in the governorate of Duhok. Mines used in the 1980s during the Iran-Iraq war are concentrated along the north-eastern border of the country. The internal conflicts illustrate a random distribution of mined areas throughout the region.

The LIS identified 3,026 distinct areas suspected of mines and/or ERW contamination. The suspected areas cover a total land surface of 776 square kilometres, which is significant concerning its relatively high intensive use by the rural population. For each affected community, a point score expressing the severity of the various mine impact was calculated. The score takes three major impact factors into account: the type of munitions presents, the livelihood and institutional areas to which mines block access, and the number of recent victims. A total of 788 of the 1,022 mine-affected communities are classified as low impact communities. Another 199 mine-affected communities are considered as medium impact and the remaining thirty-five communities are classified as high impact. The estimated size of the population living in mine-affected communities is 748,651.

The large scale of contamination and densely populated areas has resulted in many accidents. The distribution of recent victims by age and sex indicates characteristics of groups of concern. For example, 94.6% of the victims were male. Men ranging in age from fifteen to twenty-nine account for 44.8% of all victims compared to women of the same age at 1%. The most frequent categories classified by occupation involving mine accidents were: children playing, shepherds, military, farmers, and workers, accounting for 87% of all victims. The type of activity at the time of accident was closely related to the occupational breakdown, with the notable exception that nearly 30.8% of the recent victims were involved in or close to someone tampering with mines and ERW. Tampering with munitions is the most prominent cause of accidents for males and second most prominent for females. Perhaps this can be attributed to "mine hunting," whereby villagers extract metal and explosives from landmines and UXOs found lying around for recycling reasons.

A form of analysis commonly used in landmine impact surveys in other countries was applied to the Kurdistan region: regression of the number of recent victims on suspected causal factors. Many, but not all, of the clusters are near the Iranian border. Statistical analysis reveals that the distance to the nearest other community with some recent victims is a better predictor of victim numbers than the distance to the border and also shows the area of blocked access, number of communities affected and their percentages. The different areas are further disaggregated into their sub-areas. Again, pasture, crop land, and other uses of land as well as water usage areas dominate the sample. The frequency of the most common blockage type is access to grazing land at 98%. The grazing land is often alternatively used for military purposes.

This chapter also summarized past mine action activities in areas of survey and clearance operations, victim assistance, and mine awareness based on other external data sources. A summary of the Mine Advisory Group activity is also provided. The Kurdistan region has suffered from continuous conflict for more than twenty-five years, with the most recent in 2003 leading to further landmine contamination. In sum, the effect of cleared land has facilitated infrastructure and development projects including the maintenance of irrigation channels and electricity distribution.

Data collection on victims during the general survey was developed prior to the existence of a comprehensive victim assistance component within the frame of mine action program. This data captured incidence of mine/ERW accidents, basic data on victims and information on action taken immediately after the accident. No information was collected regarding the exposure of the victims to services available except where the victim was taken from the site of the accident. Most of the information gathered on victims relates to the planning and evaluation of mine awareness activities. During the landmine impact survey, limited information was collected regarding previous mine risk education activities. MAG has executed an extensive mine risk education program since 1992, focusing particularly on school teachers.

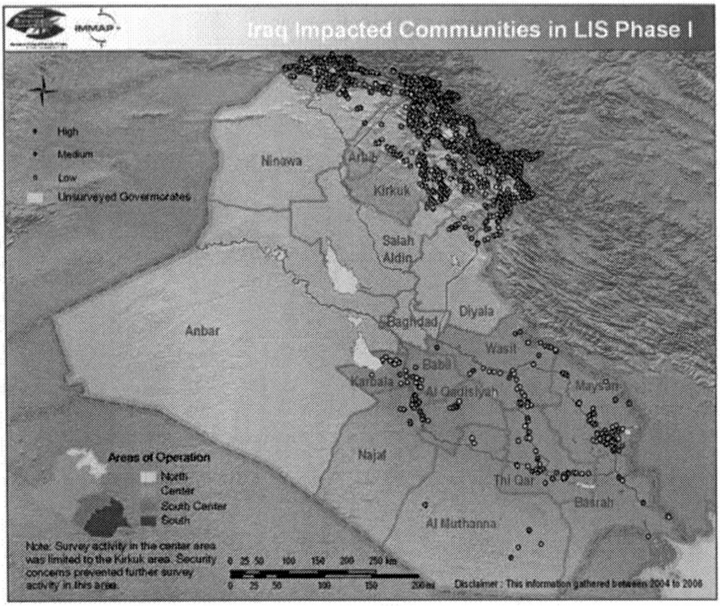

Map 6.3 Iraq Mine Affected Communities

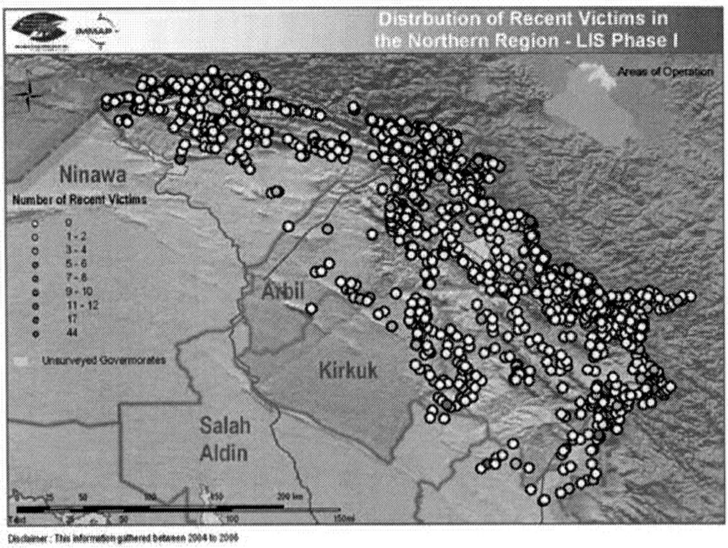

Map 6.4 Recent Victims by Community

Notes

[1] Information Management System for Mine Action which has been developed by Geneva International Center for Humanitarian Demining GICHD.
[2] The Landmine Impact Survey Report (1999–2003) implemented by the UNOPS-MAP in northern Iraq under the Oil for Food program.
[3] Emergency surgical centre in Erbil and Suleimaniya Governorates were funded by the Italian Government as a complementary process for the huge manufactured Italian mine used in the region by the previous regime against the Kurdish political parties and civilian population.
[4] A number of governments and organizations provided contributions to the LIS activities in 2006 in Iraq. The contributors included the US department of states, the European Union, and government of Italy through the UNDP.
[5] Here a community is defined as a type of settlement of a permanent or semi-permanent nature (e.g. summer village).
[6] The figures do not represent the current census suggesting a population of approximately five million living in the region with three governorates. An additional Kurdish population is living in disputed governorates and cities neighbouring the current three governorates.
[7] Recent victims refers to those involved in incidents that happened during the two year periods combined for both LIS, hence the years 1999–2000 and 2002–2003 were considered jointly.
[8] These figures are extracted from the patient database system in both emergency surgical hospitals in Erbil and Suleimaniya in 2009.
[9] Some mine-affected communities have multiple blockages, although the majority communities have none.
[10] The data is extracted partially from iMMAP-LIS and the recent IKMAC database. The estimates of the total affected population do not vary greatly among the governorates; the largest figures are from Suleimaniya and Erbil, which report approximately 266,000 compared to 216,000 from Duhok. The districts with the highest number of affected communities are Sharbazher in Suleimaniya (147), Soran in Erbil (109), Amedi in Duhok (95) and Penjwin in Suleimaniya (95). Sharbazher and Penjwin border Iran, Soran and Amedi share a border with Turkey.
[11] See http://www.sac-na.org/sac_swg.html, Accessed on July 2, 2009.
[12] The number of clearances in progress for UNOPS includes temporarily suspended minefields that are exclusively winter tasks. UNOPS implements impact surveys, technical surveys, and post clearance impact surveys, whereas other organizations are believed not to have these processes.
[13] Source: emergency surgical centres from Erbil and Suleimaniya.

CHAPTER SEVEN

FACTORS INFLUENCING MINE ACTION PROGRESS IN THE KURDISTAN REGION

Every year a large number of civilians are killed or injured due to direct contact with mines and ERW that are no longer serve military or political purposes. Mines and ERW drain limited and scarce medical resources. They hinder agricultural production and other developmental economic activities, which in turn lead to health problems, and prevent production, delivery of food and medical supplies to vulnerable populations, seriously hindering reconstruction and development efforts.

The humanitarian mine action program in the Kurdistan region, since beginning operation in 1992, has faced severe difficulties in processing its activities effectively and smoothly.[1] These difficulties can be natural or attributed to factors such as funding, manpower and lack of skilled demining personnel and foremost technology to perform the demining task. A transfer of funding and mine action activities from joint regional and international organizations to the central government responsible for the minelaying in the first place is a "blow" to the whole demining program activity.

In this chapter, we discuss the prevalence of landmines in the Kurdistan region and distribution by frequency, type and size across the three governorates of the region. In addition, we discuss natural barriers of demining operation in the Kurdistan region as well as the community, technical, economic and social issues that influence the mine action process and its outcomes. The role of education, funding and manpower is also briefly discussed.

1. Size of Mined Areas

Mined areas have different sizes ranged from less than a thousand to several million square metres. The method of clearance is partly defined by the size of mined areas. For instance, large flat areas are more suitable for mechanical clearance while small areas are more suitable for manual demining. The minefield size also influences the structure of demining teams assigned for clearance. Initially, smaller areas are targeted to provide immediate relief to mine-affected communities. Table 7.1 below indicates the size of the mined area divided by the three governorates in the region:[2]

Table 7.1. Size of Mined Areas

Governorate	Estimate of contaminated land (Square Kilometres)	Percentage of total land area contaminated
Duhok	117	1.2
Erbil	333.1	3.1
Suleimaniya	325.9	1.6
Total	776	1.9

Table 7.1 provides information about the distribution of the mined area among the three governorates of the region and their average sizes. It also provides an estimation of the total mined areas as a share of the total land area of each governorate. The amount of land included in all mined areas documented by the landmine impact survey of the Kurdistan Region ranges from 1.2% to 3.1%, with the smallest share belonging to the Suleimaniya governorate, the largest belonging to the Erbil governorate. In absolute size, the table shows that Suleimaniya has about the same total of mined area as Erbil, and each has more than twice the mined area of Duhok.

2. Mine Contaminated Land by Vegetation and Terrain

The 3,026 mined areas in the Kurdistan region identified by the LIS have different ground profiles and are covered by different types of vegetation. All the three governorates lie in the north-eastern highlands of Iraq, an area stretching from Mosul and Kirkuk toward the north and north-east where Iraq borders Turkey and Iran.

In this area, higher ground is separated by broad, undulating steppes which give way to mountains ranging from 1,000 to nearly 4,000 metres near the Iraqi international borders. These geographical terrains are critical for conducting clearance activities. Within the three northern governorates of the region, these terrain-related factors are captured during the technical survey process.

During the growing spring and early summer seasons, tall grasses and thorn bushes prevent mine detection dogs from working the ground. In order to conduct any clearance, the ground must either be burned or prepared with mechanical ground preparation tools. In other cases, steep ground or terrain obstacles prevent both dogs and mechanical preparation tools from working effectively. This detailed planning information is also captured during the technical survey process.

The LIS conducted by VVAF and iMMAP have divided the mine area data based on terrain type and vegetation into three main groups. The areas those are located:

- At lower elevations.
- On mountainsides at higher elevations, but are generally featureless.
- On mountainsides at higher elevations that include secondary features.

Table 7.2 below divides both the mined areas located at lower elevations (III), and, those at higher elevations which include secondary features (II), into two groups. In both major groups there are mined areas that are generally flat (II-a, III-a) and other mined areas that contain gullies, ridgelines and other secondary features or descriptions (II-b, III-b). The third major group includes mined areas that are featureless (no secondary features or descriptions at all) but all of which lie in higher altitude mountain areas (I).

Table 7.2. Classification of landmine areas by elevation and terrain

Location	No of Mine Areas	Total Area (Sq. m)
I. On Mountainside (Featureless)	347	59,633,124
II a On Mountainside (Flat Areas Only)	37	6,092,000
II b On Mountainside (Hills, Gullies, etc.)	1098	299,819,914
III a At Lower Elevations (Flat Land Only)	220	39,684,960
III b At Lower Elevations (Hills, Gullies, etc.)	1317	369,796,555
Totals	3019	775,026,553

3. Contaminated Land by Ordnance Type

To determine the choice of clearance methods one must also consider the logistical requirements, skill level of the demining workers and the types and distribution of munitions in the suspected contaminated areas.[3] The landmine impact survey gathers information on the generic types of munitions, i.e. different mine type such as anti-personnel (AP) mines, anti-vehicle or anti-tank (AV or AT) mines and ERW.

The majority of mine-affected communities in the Kurdistan region are affected by AP mines (93.7%), whereas only 1.3% of the communities are affected by AV mines. Fifty-two communities or 4.6% of the total are affected by mixed AP and AV mines. Four communities could not be clearly classified by munitions type during the impact survey interview.

The distribution of ordnance disposals, their type and governance location is shown in Table 7.3 below. The total area affected by both mines and UXOs comprise 54% of the total. Distribution of communities affected by ERW and UXO type is shown in Table 7.4 below. During the general survey, another 192 communities were identified as having specifically UXO stockpiles. However, these villages, due to their specific requirements, were not visited during the impact survey.[4]

Table 7.3. Distribution of ordnance, type and governorate

Ordnance Type	Erbil	Duhok	Suleimaniya	Total	Percentage
AP Landmines	646	6,98 1	6,02 2	9,46 9	7.4
Unexploded Ordnance	190	117	749	1,056	34.9
Cluster Sub Munitions	30	24	7	61	2.0
AT Landmines	26	14	57	97	3.2
Stockpiles	14	4	1	19	0.6
Other	8	1	3	12	0.4
Total Mined Area	679	709	1,636	3,024	100.0

The inhabitants of the mine-affected communities in the region are aware that most of their lands are affected by mines and ERW. This is evidenced by the available statistics as more than 97% of the 3,024 mined areas recorded in the region have AP landmines. This information is based on the information recorded from the inhabitants of the communities affected by each mined area. Almost 35% of the mined areas are said to have ERW contamination, and 2% (or thirty mined areas) to have cluster sub-munitions specifically.

Table 7.4. Affected Communities by UXO Type

ERW Type[5]	Number of Communities
Mortar	507
Bomb	181
Missile	140
Rocket	99
Grenade	67
Artillery	55
Unknown	38
Napalm[6]	12
Total	1,099[7]

Fig. 7.1 below indicates that a significant land area contains AP mines (303 sq km). This is in sharp contrast to the area containing only AV mines (2.8 sq km). In addition to that, over half of the suspected area in the region contains both landmines and ERW. This raises the urgent need for mine risk education activities in these locations, in addition to the other ERW affected communities, particularly when considering the majority of accidents related to tampering with mines and ERW.

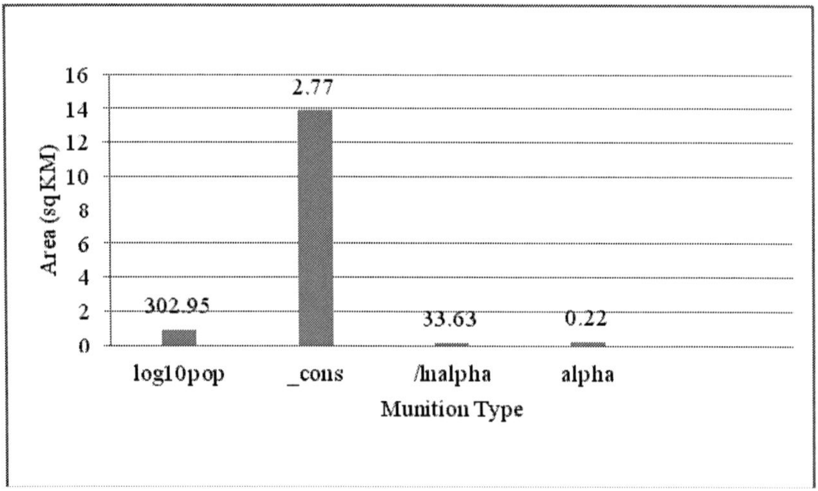

Fig. 7.1. Surface Area of Mine Affected Communities by Munitions Type

4. Influential Factors and Their Consequences on the Mine Action Process

A number of practical consequences on mine-action activities can be derived based on data presented in the earlier sections. These are divided into: general community background, technical related issues, real size of mine area, mine type, seasonal variances and growth of vegetation, and social and economic factors of the communities. Each of these factors and their impacts on mine clearance are briefly described below (Harpviken 2003).

4.1 General Community Background

There are a number of factors associated with community background that impact mine affected communities in the following ways. First, the distribution of mine affected communities throughout the three northern governorates of the Kurdistan region is unequal, which reaffirms the program's unequal allocation of clearance resources. Approximately 55% of program assets are allocated to the governorate of Suleimaniya, corresponding to 60% of the mine-affected communities located in the governorate. The UNOPS MAP resources are supplemented by those of NPA and MAG, which are for infrastructure reasons more concentrated in Suleimaniya.

Given the nearly equal distribution of mined area and resources, the current distribution of resources needs no change. High-risk communities tend to group together whereas medium and low risk communities are distributed throughout the region. This mixture in encompassing high, medium and low impact mine-affected communities in close geographic proximity suggests a distributed clustering approach to task clearance assets, not necessarily targeting high-impact areas first, but also considering sizes of the mined areas to maximize the greatest effect in the shortest clearing time.

Second, in 2006 the majority of mine-affected villages (nearly 96%) had a small population with an average of 400 villagers, yet these small rural villages account for the greatest number of victims. However, each village on average had only 0.24% of recent victims. Thus at the rural village level, one in every 1,267 villagers is at risk of becoming a mine victim. In contrast, the district level contains a mean population of approximately 19,200 inhabitants, and each district centre had on average two to three recent victims, or one in every 7,675 inhabitants at risk of becoming a victim. Overall, the inhabitants of the small villages are at three times higher risk of becoming a victim than the inhabitants of district centres. On the other hand, the area to be cleared is several times greater, thus the level of effort required for each victim in rural areas is higher than it is in urban areas.

The above facts suggest a distributed approach for both tasking clearance assets and mine risk education teams. From a geographical perspective, villages that may be grouped into clusters would result in more efficient use of resources, easing logistical support and allowing for centralized accommodation or satellite demining techniques. In terms of the magnitude of the problem, it should be noted that all the collective towns and district centres combined can be fully relieved of impact in one season by targeted clearance involving less than 20% of MAP assets.

Third, the landmine impact survey permits the identification of critical blockages, for example relating to community infrastructure, which could be separately resolved to enable mine-affected communities to have full access to such communal services. Although infrastructure blockages account for only 9%, they provide disproportionally higher benefits to the entire community, such as water or electrical services. In terms of beneficiaries, this would achieve the greatest impact. For example, in 2006 a total of sixty-nine communities claim water source blockages affecting a population of 49,270 while ninety communities claim infrastructure blockages affecting a population of 82,005. In a closer inspection of the infrastructure blockages, there are nineteen communities identified with a population greater than 500 people, including Deraluk in Duhok (25,000 people), Bani Shar in Suleimaniya (11,290 people), and Khurmal in Suleimaniya (10,000 people). It is also worth noting that it may be appropriate to give a high priority in dealing with infrastructure blockages on a temporarily basis by clearing a wide access corridor, and marking the remaining area.

Fourth, minefields claiming no impact (1,322 minefields in 572 mine-affected communities) should be targeted for early marking rather than clearance. The risk that new impacts may develop in changed circumstances, for example, with potential change in population movements and land use in a post-UN security resolution 986 context, should be considered.

Fifth, mine risk education must target children and infants as one critical group and also recognize the requirement to address the dangers to farmers and workers. The cultural lifestyle in the rural villages requires that the mine risk education teams be flexible since the working force (farmers, etc.) will almost always be tending their land or livestock during the day, whereas children can generally be found playing within the village. Since nearly half of the accidents result from tampering with either mines or UXOs, this clearly indicates an important focus area for mine risk education teams.

Sixth, in terms of victim assistance, the admission data to emergency hospitals in Erbil and Suleimaniya governorates indicates that 43% of the victims were less than eighteen years old, of which 90% were male. It is interesting to note that one sixth of these patients had an injury or accident relating to mine/UXOs. [8] Twenty five percent of the victims are permanently handicapped with another 14% remain unemployed. This reinforces the requirement for victim assistance to sustain rehabilitation and vocational training.

4.2 Technical Issues Influencing Mine Action Process

The landmine impact survey is not designed to investigate technical mine action issues in detail. This will be done during the subsequent technical survey. The impact survey findings provide valuable information in using the technical survey. Additionally, basic data gathered during the impact survey is of significant help for the subsequent overall planning of mine action.

4.3 Size of Mined Areas

The size of suspected mined areas has implications for technical survey and clearance operations. Large areas of flat land would be suitable for mine detection dogs and mechanically assisted mine clearance, while small areas are most suited for manual demining teams. The impact of large or extended minefields may also be most appropriately dealt with by breaching wide access corridors and extensive marking of remaining areas.

Distribution of the size and share of high impact areas by governorate is reported in Table 7.5 below. According to the LIS conducted by iMMAP, out of the thirty-five high impact communities, twenty-eight are located in the Governorate of Suleimaniya encompassing an area in excess of 28 square kilometres representing 81% of the governorates total high impact mine-affected area. On the other hand, seven high impact communities are located in the governorate of Erbil representing 19% of the governorate's high impact mine-affected area, encompassing nearly 7 square kilometres.

Table 7.5. Size and share of high impact areas by governorate

Governorate	Sum of High Impact Areas m^2	Percentage	No of Communities
Duhok	0	0	0
Erbil	6,648,801	18.9	7
Suleimaniya	28,440,132	81.1	28
Total	35,088,933	100.0	35

The distribution of community type affected by the grazing area and number of minefields by governorate is reported in Table 7.6 below. Consistent with earlier observations, grazing land is predominantly blocked, encompassing some 102 square kilometres or 30% of the total suspected areas. This blockage is most significant in the governorate of Suleimaniya, accounting for 542 minefields of which 70 are associated with high impact communities, 336 with medium impact communities and 136 with low impact communities. Within the Kurdistan region there are 733 minefields block grazing lands.

Table 7.6. Distribution of areas by governorate and communities affected[9]

Governorate	Community Type Affected by Grazing Area m^2				No of Minefields belonging to Community Type			
	High	Medium	Low	Total	High	Med	Low	Total
Erbil	2,432,388	5,574,686	12,257,400	20,264,474	14	66	70	150
Dahuk	0	1,914,800	854,000	2,768,800	0	28	13	41
Suleimaniya	15,967,659	40,827,801	22,207,020	79,002,480	70	336	136	542
Total	18,400,047	48,317,287	35,318,420	102,035,754	84	430	219	733

The decision of the mine action managers to address the mine problem in respect to the above mentioned aspects in highly impacted communities has additional operational and logistical advantages resulting from the concentration and clustering of impact areas.

4.4 Munitions Type

Mine type is one of the important determinant factors of clearance methods employed, influencing equipment requirements. For instance, tools, procedures and training designed for clearing areas with anti-personnel mines will differ from those employed for clearing anti-vehicle mines or ERW sites.

A total of 1,126 mine affected communities were assessed by LIS-iMMAP of which 1,030 communities cover an area of nearly 302 square kilometres and are contaminated with AP mines. In addition, twenty-five mine-affected communities contain AV mines for a total area of 2.8 square kilometres. Another fifty-three mine-affected communities contain both AP mines and AV mines, encompassing nearly 34 square kilometres. A total of 430 mine-affected communities contain both mines and ERW, comprising 184 sq km. The general survey captured 192 communities affected by ERW only. The graph (Fig. 7.2 below) provides more details on the various types of munitions.

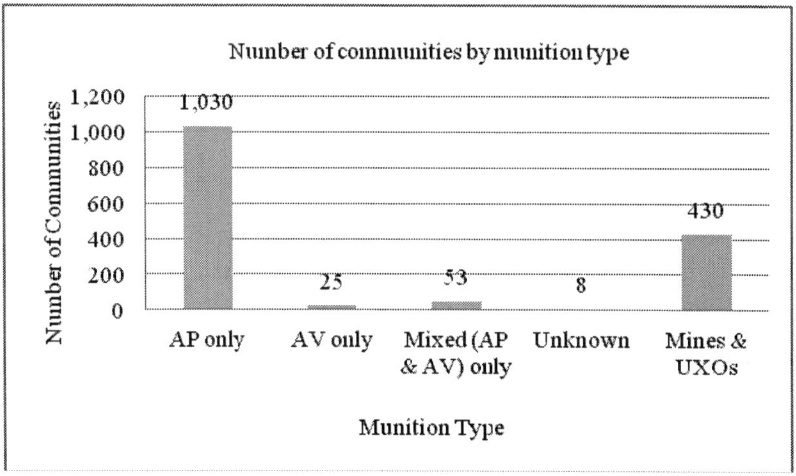

Fig. 7.2. Communities' distribution by size and munitions type

4.5 Seasonal Variances and Growth of Vegetation

Seasonal variances and vegetation influence the mine action activities. In early spring, the thorns prevent dogs from working effectively, thus, a mechanical ground preparation tool must remove the thorns. Later in the season, the heavy vegetation must be addressed in a similar manner.

Mining of built up areas is not a major problem in the Kurdistan region. This will eliminate the need to invest in equipment designed for built up area clearance and extensive training of mine action staff for this purpose. However, many former rural communities are reduced to a state of rubble. These are often contaminated in many cases with landmines and ERW. In spite of this, many communities have re-constructed homes adjacent to the rubble. This leads to demand for armoured mechanical devices such as front end loaders.

4.6 Social and Economic Factors

Recent research unambiguously concludes that social and economic factors to a large extent affect the behaviour of risk taking including the mines and ERW risk. A report published by landmine action,[10] in 2003 states that "decisions to take risks with Mines and ERW are not only a balance between economics and fear, but are also conditioned by important social factors, including a sense of social responsibility." In addition to the responsibility for the families' living conditions, in poor communities, mine, ERW and other military debris have a high value because of the value of the scrap metal, the result being the utility of the explosives attracts victims.

Furthermore, the data in the report show that while the great majority of mine and ERW victims are men, children make up a significantly large proportion. The size and shape of the munitions, particularly the bright colours of certain sub-munitions, make them attractive to children. In recent years, the mines used in relation with conflict at southern Lebanon exclusively contained this type of ordinances.

The mine awareness program would therefore have to take into consideration social factors such as poverty, gender and age in reducing the risks and negative outcomes. Measures are required to be effective in different contexts, for example among refugees, internally displaced people, resident populations, and nomadic groups. The already extensive landmine/ERW awareness education efforts should serve as a basis for further deliberations on this issue.

5. Summary of Factors Influencing Mine Action Progress in the Kurdistan Region

In the Kurdistan region every year a large numbers of civilians are killed or injured by mines and ERW, creating a drain on scarce medical resources, and effectively hindering all rural economic activities. These in turn lead to the prevention of the production of food, causing a shortage of food supply and severe health problems. The impact is strongest on the most vulnerable rural populations, and has hampered reconstruction and development efforts to bring peace and normality to the communities.

The humanitarian mine action program in the Kurdistan region has faced difficulties in effective operation. These difficulties can be of natural characteristics or attributed to factors such as funding, manpower and lack of skilled demining personnel and foremost the availability of technology to perform effective demining. A transfer of funding, planning and implementation of mine-action activities from the joint regional and international organization to the central government is a "blow" to the whole demining program activity as the latter was responsible for the minelaying in the first place, and as such should not trusted to fund, plan, implement and confirm demining activities. It will be hard for the suffering rural population to entrust the central agencies to take responsibility for such a vital factor. More appropriate would be that the central government takes its responsibility for the clearing by financing a third party to undertake it and to provide rehabilitation and training programs, compensating the victims and their families.

In this chapter, we have discussed the natural barriers of demining operation in the Kurdistan region as well as the technical issues that influence the mine action process, its progress and outcomes. The practical consequences from influential factors are many and must be taken into account in appropriate stages of the task. These are divided into general community backgrounds, technical issues, the size of mine area, mine type, seasonal variations and growth of vegetation, and social and economic factors. Each of these factors and their possible impacts on the mine clearance activities has been briefly described above and will be further discussed in the following chapters of this book.

Education seems to play an important role in the context of mines and reduction of their threat and impact. It is desired that mine awareness programs take into consideration social factors such as poverty, gender and age. Economics, gender and age are among the factors found to have great significance in different locations. Measures are required to be effective and correspond to the heterogeneous characteristics and needs of different groups and societies such as refugees, internally displaced people, resident populations and nomadic groups.

Notes

[1] Mine action in the three Northern governorates of Iraq (2003), Situation Report, UN Office for Project Services, available at http://reliefweb.int/node/104651, Accessed on Aug, 21, 2009.
[2] The table is taken from the LIS conducted by iMMAP in 2006.
[3] The United Nations Mine Action Services (UNMAS), *International Mine Action Standards* (*IMAS*), NewYork: UNMAS, 2001, available at http://www.mineactionstandards.org/imas.htm, Accessed on Dec, 15, 2009
[4] These sites represent tasks suitable for Explosive Ordnance Disposal (EOD) teams.
[5] Mortars, bombs and missiles are the primary types of ERW and UXOs impacting mine affected communities.
[6] Information on Napalm: This is an Incendiary bomb with a TNT buster. The filler constitutes about 50 to 75 percent of the total weight of the bomb. Some filler are thermite, thermite balls, naphtha and thermite, phosphorus and benzene-and-oil, and thermite with an "electron" case. The body is gray with a red band. Incendiary bombs are made of sheet iron or magnesium alloy. The sections are welded together. The "electron" case is an alloy of manganese, copper, zinc, magnesium, and aluminium. Bombs using "electron" cases use thermite fillers instead of thermite because thermite produces more heat; this heat ignites the case. The Napalm is of diameter 279.4 mm, length 982.98. Weight: 114 kg, NEQ 28.4 kg.
[7] The actual number of communities affected by UXO is 192. Multiple occurrences account for higher count.
[8] Clearly the Emergency Centers are accepting other patients as well and the greatest form of treatment accounting for 65% is for burn injuries resulting from hot water or kerosene
[9] Totals grazing for communities differ from overall high, medium, and low mine affected communities, since the table only pertains to those mine-affected communities affected by blocked grazing land and these totals reflect multiple occurrences.
[10] Explosive Remnants of War – Unexploded ordnance and post-conflict communities, published in March 2002 by Landmine Action.

CHAPTER EIGHT

HISTORICAL REVIEW OF INFORMATION MANAGEMENT SYSTEM FOR MINE ACTION IN THE KURDISTAN REGION

From the beginning of the mine action program in the Kurdistan region of Iraq in 1999, the UNOPS established a comprehensive information management system. The system was designed to allow the management of all mine related data. The initial step was to import all minefield information records from the Mine Advisory Group (MAG) database. The Mine Advisory Group MAG was involved in mine action activity in the region since 1992. Its major activity was marking all the minefields for future clearance. The new system was called the UNOPS Mine Action Information System (UNOPS-MAIS).

The Kurdistan region's mine action information database was originally developed and designed to store level one survey information. Level one survey information is aimed at describing the overall mine and ERW problem in the region. Although the database has fulfilled the initial requirements, however, it has not stayed abreast with later developments in the mine action sector. A number of requirements suggested that the database should be upgraded to fulfil such conditions. The requirements include those related to the prioritization of mine action activities on the basis of their impact on communities, changes in the UN mine action standards, strategic planning and development of various possible scenarios for mine action programs and the ideas of "mine free" and "impact free" conditions. In order for the database to meet the abovementioned requirements, it had to be upgraded to incorporate the current trends in mine action.

In order to upgrade the database to keep up with new trends and requirements, a proper software engineering process was followed. An initial assessment of the existing database functionality and its application in mine action activities has been defined. After defining this functionality, an assessment was done to determine which additional functionality would be needed to implement and integrate information aspects to adhere to the new international mine action standards for humanitarian mine clearance operations.

The assessment led to recognize that there would be a need to verify the quality of the existing data, correct or complete missing fields, modify the structure of the database to include new fields for blockage/impact data, and collect the corresponding data. Based on the available data in the original database, certain factors and parameters were used to prioritize the communities for the initial follow up on impact survey in the Kurdistan region.

The factors needed to be identified for prioritizing ERW and mine-affected communities had to measure certain impacts. This impact had to be measured against the relief that mine action activities would provide the targeted communities. In order to define this impact, certain blockage factors have been defined in general and of a more country-specific character. Therefore, country-specific blockages such as access denied to bathing water, hunting fields, collecting materials for shelter, and collecting wild plants were defined and considered.

In conclusion, the process described above restructuring the database, verifying and correcting data and collecting additional data on the impact of landmines on communities through a follow-up survey allowed the Kurdistan region's mine action program to enhance its strategic management and priority setting capability. In addition, it allowed for implementation of the new international mine action standards, while continuing to service the current information management needs of the program.

In this chapter, we provide an introduction to the UNOPS mine action information system in Section 2. Section 3 is about the information management system for mine action. Section 4 is about the integration of Mine Advisory Group Database (MAG-DB) with an information management system for Mine Action IMSMA. In Section 5 we discuss the system for reporting data from MAG to IKMAC. The UNOPS mine action program information management policy is discussed in Section 6. Section 7 deals with information custodian tasks. Section 8 and 9 are about responsibilities and information management throughout its life cycle. In Section 10, the strategic direction for the future is investigated and its implementation is found in Section 11. The final section 12 is a summary of information management systems for mine action in Kurdistan Region.

1. The UNOPS Mine Action Information System

The UNOPS-MAIS was an information system developed by the information section of the MAP. From January 2002 to August 2002, a US

$60,000 investment was made by UNOPS-MAU to hire a database consultant to streamline the demining module and develop mine risk education (MRE) and mine victim assistance (MVA) modules. The MAIS was a fully integrated database with modules for demining, MRE and MVA that are cross-linked and can support the integrated approach described in chapter seven of this book. The user can generate a very large number of reports in order to obtain management information from the database.

The UNOPS-MAIS was distributed throughout twenty-five locations in the region, including the Kurdistan organization of mine awareness (MRE-NGO-KOMA), mine victim assistance (MVA-NGO), emergency and the related prosthetic limb centres, surgical centres, outreach posts and first aid posts, demining NGOs such as Aras Demining Organization (ADO), Pirmam Demining Organization (PDO) and Tiroj Demining Organization (TDO), Demining contractors such as S3 and MECHEM and UNOPS Sector Offices in Duhok, Soran Fort and Suse Fort. Demining data is exported on a weekly basis from all locations and imported at the to the head office, and MRE and MVA data on a monthly basis. A list of the mine action organization in the Kurdistan region is shown in Table 8.1 below.

Moreover, the information section managed the development of certain management tools such as PATHWAY and the spatial management information tool (SUMMIT), in coordination with the Cranfield (Mine Action) University in the United Kingdom (CMA). PATHWAY is a strategic planning tool that creates long-term scenarios based on the information in the UNOPS-MAIS. This tool automates the process of creating "what-if" scenarios based on the criteria decided by the user. It takes into consideration variables such as: area to be cleared, resources used, clearance rates, clearance methods, and operation timeframe. SUMMIT is a geographical information system (GIS) tool that allows the user to display and digitize all information in the UNOPS-MAIS on maps in various ways. It produces geographical maps in a very simple procedure, and displays management information spatially according to user-defined criteria.

Table 8.1. Mine Action Organizations in Kurdistan Region

Organization Name	Short Name
Tiroj Demining Organization	TDO
Kurdistan Organization for Mine Awareness	KOMA
DDG (Danish Demining Group)	DDG
Handicap International (France)	HIFr
General director of Mine Action –Suleimaniya	GDMA
IFRC (International Federation of Red Cross and Red Crescent Societies)	IFRC
Permam Demining organization	PDO
Iraqi Kurdistan Mine Action Agency	IKMAA
Bawajy Demining Organization	BDO
Norwegian People Aid	NPA
United Nations Development Program	UNDP
Vietnam Veterans American Foundation	VVAF
Mine Advisory Group	MAG
Zozik Company Mechanical	Zozik
Information Management for Mine Action Programme	iMMAP

2. The Information Management System for Mine Action

There was a need to integrate all mine action information systems based on national level in the countries performing mine action activities. Based on such desire, the Information Management System for Mine Action (IMSMA) has been developed for the United Nation Mine Action Program by the Geneva Intention Centre for Humanitarian Demining (GICHD)[1] with technical support provided by the Swiss Federal Institute of Technology. It was suggested that the IMSMA will be used in national mine action centres, by organizations performing field assessments and surveys, and by other organizations and agencies engaged in mine action at field level. The IMSMA is a combination of database and GIS components, based on International Mine Action Standard (IMAS) to be used worldwide within mine action. IMSMA is a tool for planning, managing, and reporting, and its field model is, according to GICHD, capable of data aggregation provided by the field models giving information for mine action planning, international coordination and decision making at the strategic level (see Fig. 8.1).

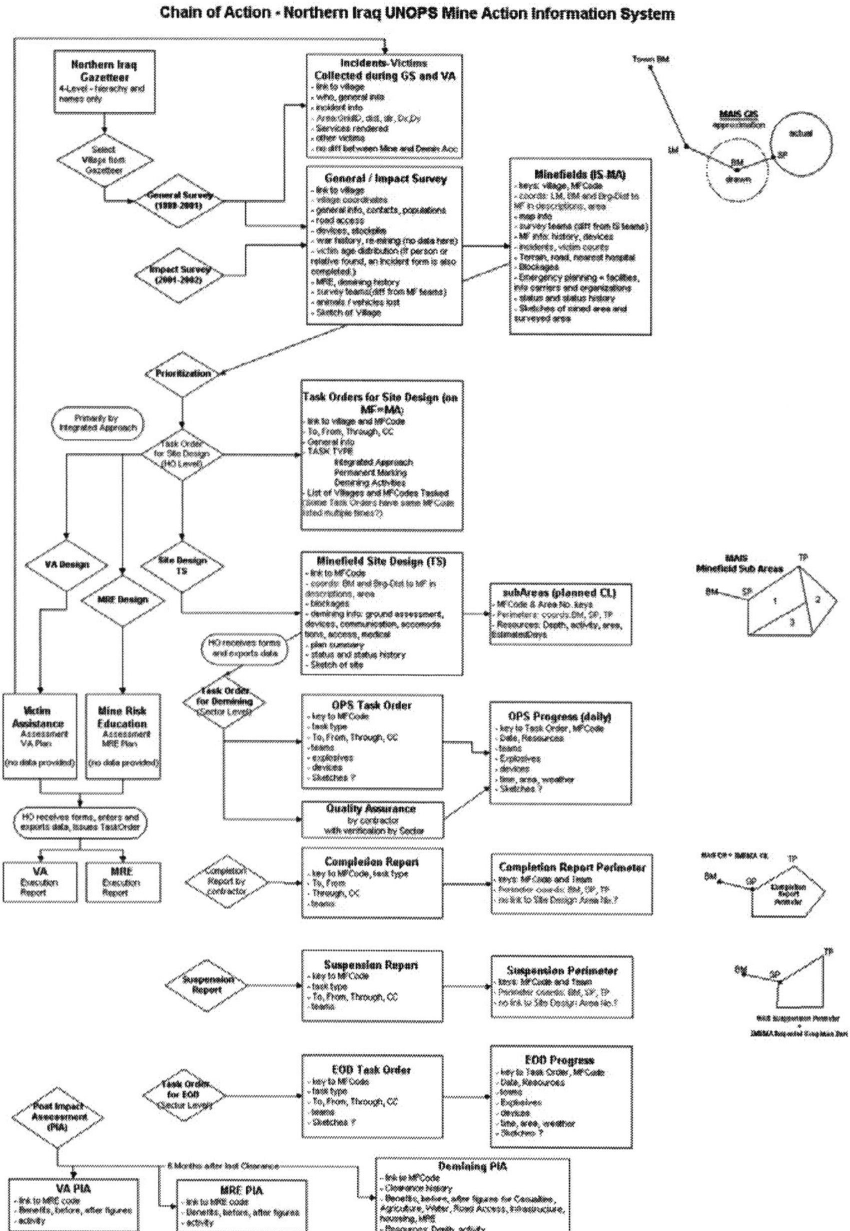

Fig. 8.1[2] Integration of Mine Advisory Group Database (MAG-DB) with IMSMA

By definition, the IMSMA will standardize the way in which mine action information is held at the national mine action centres, and the way in which reports and returns are provided to the UN headquarters. Later versions of the IMSMA field module will provide managers at national level with decision support tools, providing a common approach to the management of information based on best practice. It is expected that the module should lead to consistent and transparent management decisions and properly orchestrated collective action, without constraining national initiatives.

The advantage of IMSMA compared with its predecessors can be summarized in the following points:

- Provides improved operational capabilities related to Operational (Demining), Mine Awareness (MA), Mine Assistance Victims and Fields Support Services activities, and development for planning, decision making, information sharing and information overview.
- Enables historical tracking of the components mentioned above and their development over time.
- Generates standard statistics and reports regarding all the components.
- Leads to cost reduction due to better allocation of resources, better overview for projects, better progress overview of mine action activities and better and detailed overview of contamination.

3. Integration of Mine Advisory Group Database (MAG-DB) with IMSMA[3]

3.1 Historical Background

Since the MAG's involvement in mine action in the Kurdistan region from the early 1990s, the organization has achieved widespread survey and clearance work and built up an extensive database on this work. In 1999, the MAG database was used as the baseline of the UNOPS mine action database. About 2,300 minefield records were imported into this newly developed database.

Unfortunately, this was the last and only time such an exchange of data occurred and the two mine action programs have continued to operate in the same areas for five years, in isolation, and without any formal exchange of data (for mainly political reasons during the oil-for-food and previous regime era).

After the initial exchange of data in 1999, the UNOPS program has undertaken several surveys, including a general survey, a local version of a level 1 survey (impact survey), and about four hundred technical surveys. This resulted in a comprehensive database that included the 2,300 minefield records originally obtained from MAG.

During the last year of the UNOPS program, the data in the UNOPS database was converted into IMSMA format. In January 2004, the Kurdistan region started using IMSMA in an operational mode in practice. The additional functionality of the UNOPS database was retained in the form of add-on modules for IMSMA. Although the data conversion was comprehensive, as could be expected, it was not possible to convert all the data between the two systems perfectly because of the different functionality and data structures of the two systems. This reason that the UNOPS impact survey was not conducted according to international standards (due to the political isolation of the program) have contributed to the decision to re-survey the three northern governorates as part of a nationwide Iraq landmine impact survey (ILIS) which was led by VVAF. The ILIS is now in full swing in the northern governorates and, using IMSMA, systematically replaces information collected by the UNOPS impact survey.

Considering the history of survey and databases in the Kurdistan region, it is clear that a wealth of information was collected during the last 10 to 12 years, but that not all of it is captured in a single system. There was, however, a considerable overlap between the MAG and the IKMAC-IMSMA databases. The current system, ILIS, is providing data of excellent quality, coverage and comprehensiveness, especially at the community level. In this work, we have utilized this database for our analysis of the landmine problem and its socio-economic impacts in the Kurdistan region.

In the spirit of cooperation, the two field models have been combined to enable the IKMAC to have one comprehensive mine action database for the Kurdistan region that forms part of the greater national IMSMA database for Iraq. By combining resources, all demining organizations are able to plan and manage their operations from the same database, and an optimum level of data sharing can be reached.

3.2 A Comparison Study between MAG-DB and IMSMA

There were approximately 2,900 minefield records in the MAG database. Of these, around 2,300 have corresponding records in the IMSMA. The 600 remaining records were partly new records that captured below the “green line” (the zone identified by coalition forces after the 2003 invasion into Iraq), and partly minefields merged by UNOPS and thus removed from the database. A small part is classified as those deemed not to be a minefield by villagers, and thus removed from the database after being signed off by the village heads.

In addition to the minefield records, the MAG had approximately 600 completion reports that are not in electronic format. The 2,900 records have already been converted into IMSMA by the MAG database personnel. The 600 completion reports were also rendered into the IMSMA later. It is necessary to look at the IMSMA data flow and chain of action to understand how to practically incorporate the MAG data into the IMSMA.

As shown in Fig. 8.1 above, the normal chain of action in the IMSMA environment is to start by conducting an impact survey on mine-contaminated villages. The impact survey then identifies mined areas and records specific blockage and victim data entries for them.

A mined area (or area or hazard) is a general area that, in some way, presents a risk to the community and is not defined as a minefield by the impact survey. Because the surveyors are not technical demining staff, they only make estimates of size and collect very limited information on the found ordnance.

The next process in the chain is the technical survey. During this process, trained technical surveyors and demining personnel define the perimeter of the minefield and collect detailed information on ordnance and devices, ground profile, and confirm the blockage factors. The result of a technical survey is a clearly defined minefield. It is also possible that one mined area could actually be two or more minefields.

One of the main differences between the MAG and the IMSMA databases (one that had an impact on the data transfer), is that the MAG database did not follow the same chain of events in order to reach the final step of creating a minefield object in the database. Fig. 8.2 below shows the mine action chain of events in the IMSMA. The information inside the red square is collected by the ILIS, while the information inside the blue square is collected by demining organizations.

The MAG database was not designed to record the impact survey data. It was focused on demining operations; therefore it records only the last object in the IMSMA chain of action, namely the minefield record.

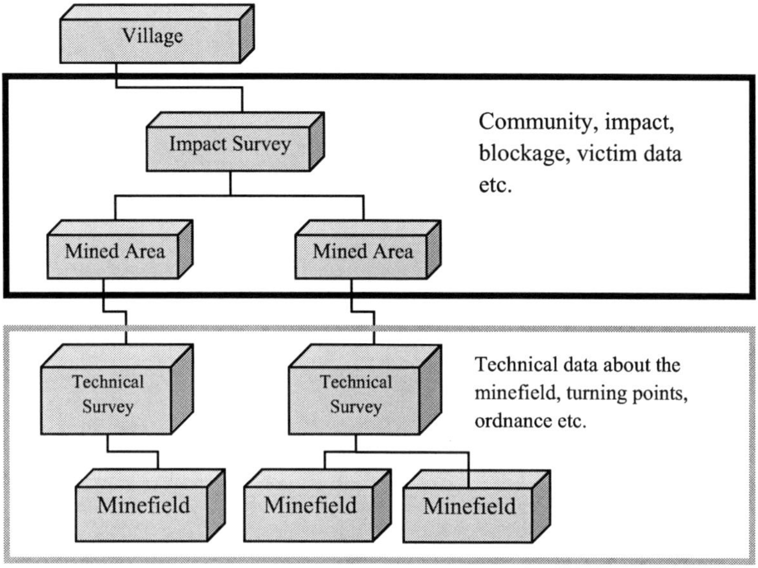

Fig. 8.2. The mine action chain of events in IMSMA.[4]

During the conversion of the MAG data to the IMSMA format, it was necessary to create the whole chain of events inside the IMSMA to create the minefield records. This means that the MAG-IMSMA database has impact survey, mined area, and technical survey records that are largely empty except for location data, and serve only as the necessary links required by the IMSMA chain of events rules to the Minefield records. The main reason for creating this "pseudo" chain of events is because the MAG has turning points (minefield perimeters) for most of their records, and the only place to store these perimeters in the IMSMA is in the minefield object. Therefore, it was necessary to create the whole chain of events with empty records to allow the creation of minefields in the IMSMA.

The MAG also has a few data fields of additional data that is not standard in the IMSMA, or that do not function as required in the IMSMA, two examples being the status and the minefield/task code. User defined fields were used to capture these data.

3.3 Updating IKMAC-IMSMA with MAG-DB

There were two major components in the process of updating the IMSMA database with MAG data. First, a shift in the way that the IMSMA is operated in Iraqi Kurdistan Mine Action (IKMAC) with regards to technical surveys and minefields data, and second, the comparison of data and the technical transfer and/or updating of data from the MAG-IMSMA database to the IKMAC-IMSMA database.

The IMSMA was operated strictly in the way it was intended to be used. This involved the full mine action chain of events (see Fig. 8.1) from impact survey, mined area, technical survey, and then to minefield. This indicates that there are only about four hundred of the almost 3,800 mined areas that have undergone a technical survey and thus have a perimeter and all related technical information in a minefield record.

The IMSMA was designed to follow a certain demining process as specified by IMAS. Therefore, it has very clear boundaries for each process, and expects every process to be completed fully before the next in the chain of action. However, this is a scenario that only really happens on paper and in an ideal world. In the field, there are several other factors that influence the way elements of field work are conducted.

Furthermore, it is in the interest of all demining organizations to have access to the best data available, and it would be wrong to deny them because the information system is not flexible. By making use of the IMSMA functionality slightly differently from the way it was intended, it will be possible to provide everyone with all available data.

In order to incorporate the additional MAG data into the IMSMA, it was necessary to stray slightly from the traditional IMSMA way of operation, and allow real processes happening on the ground to be accommodated inside the IMSMA. This has allowed the IKMAC and demining organizations to have full access to all available data. In practical terms, this means that it was necessary to make use of the technical survey and minefield modules in the IMSMA to record partial technical survey and minefield data. Thus, by creating a technical survey for a mined area, it is possible to create a minefield record and record the perimeter data available from MAG.

By using the system in this way, it was not only the MAG data that are accommodated, but also was possible to convert some dangerous areas to minefields and manage them properly with regards to tasking, progress reports and completion of reports.

Even though the mined areas affected by this data transfer have technical surveys and minefields related to them, it had no effect on the way they are tasked for clearance and how information is gathered. A clearance task was still preceded by a technical survey and a minefield report. The only difference was that the tasked clearance organization had the additional information such as minefield perimeters and area descriptions as recorded by MAG at their disposal when moving out to do the technical survey. Instead of creating a new Technical Survey process, a current technical survey was re-tasked and the information was updated.

Thus, by changing the way the IMSMA is used, and by using the technical survey and minefield modules to record partial but valuable data from MAG, everyone stands to benefit.

The actual transfer of the data takes place such that the technical transfer of data has been made considerably easier and less time consuming by MAG because they already transferred all their data into the IMSMA format. The main task would be to make a comparison between the MAG and the IMSMA data, update the IMSMA data where appropriate and transfer the additional data fields to the IKMAC-IMSMA database. This process required the close cooperation between MAG and IKMAC-IMSMA personnel.

It is necessary to mention a number of specific points to be considered during the transfer/update process. These include the following:

- The creation of Minefields included the transfer of all related information from the Mined Areas.
- Special care has been taken with the 600 records that were not in the IMSMA. They have been linked to the correct mined areas and where no mined area is available, the impact survey and related mined area records have been created.
- Where two mined areas were previously merged by UNOPS, it was possible that they could be separated into two separate minefields again, as they were originally recorded in the MAG data.
- All MAG completion reports have been entered into the IMSMA.
- The status rules in the IMSMA have been followed to avoid dependency errors.
- Specifically, the DC (demarcated) status of MAG records should be updated in the minefield record in the appropriate space as provided in the IMSMA (marking).
- The IKMAC database managers were already considering the creation of an add-on priority calculation tool that used only the blockage factors from mined areas and minefields, because when the minefield record is created, the mined area status is set to “closed” and no longer contributes to the following impact calculations.

4. The Data Reporting System from MAG to IKMAC

This section explains the system of data reporting from MAG to IKMAC, and tasking procedures from IKMAC to MAG. The migration of the MAG database to IKMAC-IMSMA has enabled both parties to use the same database for planning and managing mine action operations. IKMAC and the National Mine Action Authority (NMAA) have all the additional MAG data available to provide better information to demining organizations. All information that MAG entered into this new system is automatically reflected in the NMAA Baghdad IMSMA database each time database synchronization takes place between Erbil and Baghdad. Moreover, MAG has the full Iraq IMSMA database with impact survey data available to plan future operations.

After the migration of the MAG data and providing MAG with a full copy of the IKMAC-IMSMA database, MAG will do all future data collection and reporting on IMSMA forms. This includes (where no impact survey has been done) impact survey and mined area forms, as well as technical survey, minefield and completion survey forms. It is up to MAG to use the IMSMA progress reports or the IMSMA demining progress add-on database. If the add-on database is used, the progress and completion information will be created by the database.

Two basic options as listed below for reporting data to IKMAC were made possible after the data migration. At this stage, the first option was preferred by IKMAC:

- MAG to fill in the IMSMA forms in the operation field and hand them over to the IKMAC, where it will be entered by data entry personnel. The IKMAC then provides an update of the full database every one or two weeks (as is the case with all the local NGOs).
- MAG can use their own IMSMA data entry personnel and enter all forms at MAG location. MAG will then provide an electronic export of data to the IKMAC every one or two weeks, upon which the IKMAC will only carry out quality assurance (QA) on the data before importing it into the IKMAC-IMSMA database.

Regardless of the preferred option, the IKMAC is responsible for keeping an archive of all forms entered into the IMSMA. This could be duplicated at MAG location, but it is not necessary and will only take up extra space. With regards to tasking, the options are similar to those for reporting outlined above, again with the first preferred option at this stage:

- After coordinating a task list with the IKMAC, MAG can fill in the IMSMA task forms and hand them over to the IKMAC where the task will be created in IMSMA, assigned a task number and the task dossier handed over to MAG.
- After coordinating a task list with the IKMAC, the MAG-IMSMA database manager can create the task dossier in the IMSMA. MAG will then provide an electronic export as part of the regular data export to IKMAC, upon which the IKMAC will have the MAG tasks in the IMSMA.

5. A Unique Opportunity for Pilot Testing of New Web-Based Developments

One of the objectives of information management is to provide the end user with good quality information in a most efficient way. This implies that the turnaround time of the cycle of data collection, input, storage, analysis and output should be as short as possible. By making use of the first option above, that of electronic transfer of the IMSMA data from MAG to IKMAC, it will already be a step in the right direction in terms of making data accessible to the end user and the planner, quicker and with a smaller amount of manual work in between.

In this era of interconnectivity and the internet, we have even more opportunities to make data reporting, analysis and output faster and more readily available to the end user. By making use of a web-based system, it is possible for the end user to have real-time information available as soon as data is entered into the system.

The IKMAC has already invested time in developing web based data entry directly to the IMSMA database. The next step would be to have all the analysis and reporting functionality of IMSMA available in the same web-based system. This will completely eliminate turnaround time, and as soon as the data entry person enters a field report, it will be available for planning.

The level of expertise and technological skills in MAG provides IKMAC with an ideal opportunity to pilot test this system. It is therefore strongly recommended that IKMAC and MAG work together to test this system in an operational environment. It is often said that information systems are developed without the input from the end user in the field. By having a mine action organization with a lot of experience and skills taking part in the development and testing of this system, it will certainly benefit all future users.

Through good coordination and cooperation between MAG and IKMAC it was possible to merge the two information systems to benefit both organizations. It is in the interest of both organizations to share all mine action data, and to regularly report demining progress. By doing the migration of additional MAG data to IKMAC-IMSMA and adopting IMSMA as a mine action information system in MAG, all parties will be working from the same baseline.

There has been no more confusion when referring to MAG minefields or IMSMA minefields, MAG tasks or IMSMA tasks, and no more confusion about how many minefields have actually been cleared. All reporting will be in the same format and all MAG progress will be immediately available from IMSMA, whether the query is made in the cities of Erbil, Baghdad or Basrah.

6. The UNOPS Mine Action Program Information Management Policy

This section provides information about the UNPOPS mine action program's information management policy, its objectives and statements, as well as managing the UNOPS MAP information system. Information must be managed effectively to support and enhance the UNOPS-MAP's ability to deliver its information, programs and services in an efficient manner. Information management is a process to ensure that quality information is available to decision makers in a timely manner, and it facilitates documenting the decision making process.

The information management policy provides the UNPP MAP with a framework for managing information in a comprehensive and strategic manner. Managing information-based resources is as critical as managing human and financial resources. Managing the collection, creation, organization, and retrieval of information is vital to supporting effective decision making and efficient delivery of programs and services to end users.

Information management is the coordinated management of an organization's information based resources including its information holdings and investment in technology. It implies planning, directing and controlling all of the organization's information based resources to meet corporate goals, and to deliver required programs and services. It is a consequence of the premise that an organization's information holdings and investments in information technology are valuable resources and as such are critical factors in the achievements of its objectives.

The purpose of this policy is to ensure that information under the control of the UNOPS-MAP is effectively managed throughout its lifecycle. In order to achieve the expected result, the UNOPS-MAP must manage information in a manner that supports the program's goals and activities, its delivery of information and services to the people of the Kurdistan region and its commitment to local capacity building. Thus, the application policy is applicable to the UNOPS-MAP.

The UNOPS-MAP collects information from three distinct activities: demining, mine risk education and mine victim assistance. The information officer is responsible for the overall management of information and the UNOPS Mine Action Information System (MAIS) under the direction of the chief of planning. The information office is responsible for the development and maintenance of the MAIS and provides the tools for data capture, record management, and retrieval of information.

All three section heads (mine risk education [MRE], mine victims assistance [MVA], and operations [OPS]) have the responsibility for identifying and managing the information in their area. Managers at all levels of the organization must ensure information management activities align with the goals of their work area and facilitate the integrity, availability and confidentiality of information.

7. The Information Custodians Tasks

This section covers the information custodians, assignment of responsibilities, issues of integrity, confidentiality and accessibility to information, as well as reducing unnecessary efforts made in the process of data gathering and management.

Managers should ensure that all information in their area has an identified custodian. Responsibility for some information can be clear and defined, as it is defined as part of some job descriptions. However, responsibility is not always so clear, and managers should clarify responsibility for information. The identified information custodian is responsible for the integrity, accessibility and disposal of that information.

The information custodians are responsible for maintaining the integrity of their information so that it is appropriate, correct, and up-to-date. They should ensure the following:

- Information is correct before relying on it.
- Information is kept up to date.
- Information cannot be manipulated by unauthorized people.

The information custodians are primarily responsible for ensuring that confidential information is kept and is disclosed appropriately, and should ensure their information is available only to the appropriate people or organizations, safeguarding it against those involved in mine laying that targets the civilian population.

All managers should ensure that any information management activities are justifiable and do not involve unnecessary or duplicated effort.

8. Responsibilities

Responsibilities involve managers, information custodians, executive management and information officers. Each of these levels of organization and their interrelationships is discussed in this section.

All managers are responsible for ensuring that the information in their area is appropriately managed, either by themselves or by custodians. Where information is not assigned to a custodian, the manager is the default custodian.

The information custodians are the prime responsible persons for ensuring that information in their care is:

- Secure, so that its integrity, accessibility, and confidentiality are maintained. The information office can assist in ensuring the security of computerized information.
- Appropriately made available, in consultation with their manager.
- Appropriately stored, archived or disposed when it is no longer regarded as new.

Executive management is responsible for providing adequate direction and guidance for information management issues. Finally, the information office is responsible for providing the management regime and support services for UNOPS MAP information resources.

9. Managing Information throughout its Lifecycle

All information under the control of UNOPS MAP must be managed and protected effectively regardless of medium or format. In order to ensure an effective and efficient information management, the sections must ensure the followings are achieved:

(a) Make plans to:

- Integrate information management into current Standard Operations' Procedures (SOPs).
- Include human and financial resources, information needs, and information management requirements in strategic planning.
- Optimize the use of existing information and make plans for its use beyond immediate needs.
- Maintain the authenticity, integrity and currency of information for as long as the information is required.

(b) Collect, create, receive, and/or capture information in a manner that ensures it is available, understandable and useable for as long as required.

(c) Organize, retrieve, use, make accessible and disseminate information.

(d) Store, maintain, protect and preserve information.

(e) Dispose of or preserve information to:
- Eliminate unnecessary information.
- Ensure the preservation of historical information and the reliable access to information of enduring value.

10. Strategic Direction for the Future

Three major objectives for enhancing services through the innovative use of information and technology are identified: investing strategically, building partnerships, and distributing computer power to managers and staff, each of which is described below.

10.1 Investing Strategically

There are many opportunities to improve mine action operations through investments in information technology. Some of these investment opportunities are new applications while others replace older systems. Many cannot be funded from existing budgets, and thus would require new sources. In the present climate of restraint, it is not possible to fund all proposed projects. Therefore, the right investment choices must be made to optimize opportunities to improve the program. The major information technology projects have to be evaluated based on criteria such as:

- Return on investment (reduce expenditure and improve service availability and quality).
- Enhancing competitiveness in the market.
- Increase in feasibility and risk reduction.
- Financial arrangements and sources of funds.
- Information sharing with other programs.

The following strategies and related initiatives should be incorporated the following aspects:

(a) Identifying projects for potential investment as follows:
- Identify opportunities for new applications/developments.
- Review the current application/structure and hardware base to ascertain what is obsolete or what has excessive operating or maintenance costs.
- Design strategies to modernize obsolete systems/structures, and to free up resources dedicated to their maintenance

(b) Evaluate, select, and fund information technology projects as follows:
- Rank unfunded investment proposals against the established criteria.
- Establish a funding strategy.
- Select projects for investment.

(c) Promote sharing to avoid duplication and minimize funding as follows:
- Examine proposals for new projects to identify opportunities to share existing information, technology, applications and facilities
- Encourage other programs to identify and share common systems and information/data.

10.2 Building Partnerships

Partnerships with other eligible programs and alliances with the private sector that do not jeopardize safety and security of individuals can be beneficial. Partnerships are formed to overcome inherent weakness or overcome budgetary or resource limitations, and should result in mutual benefit. Partnerships can range from simple exchange of personnel into joint ventures to outsourcing an entire operation. Opportunities to share applications and facilities and to adopt common standards can reduce costs to all parties as well as improve service to the programs. When planning to share data with other programs, privacy and security requirements must be respected.

10.3 Distribute Computing Power to Managers and Their Staff

The decentralization of computing power into the desktop is shifting the knowledge of responsibility and control of computers and information away from experts to managers and their staff. The growing trend away from "technology push" towards "user pull" in the drive for change should be encouraged. The past emphasis on improving production through automation has shifted much of the responsibilities for designing business processes from the program manager to the information systems function. The current emphasis on improving service and program delivery allows program managers to re-assume prime responsibility. If senior managers are to lead these changes rather than react to them, they must be provided with and be willing to accept increased education and training in the use of enabling technology and in the management of information resources.

Similarly, junior managers and their staff must also receive training so that they can fully manage their information resources and better apply the power of their computers to optimize the way they do their jobs. Furthermore, it must be realized that significant resources have to be allocated to support the extensive information, software, hardware, and networks that will be implemented.

Enhancing employees' knowledge and skills will increase productivity, improve service and, at the same time, make the program a more attractive place to work. This will support the recruitment and retention of competent staff.

The following strategies and related initiatives will help to achieve the objectives that have been discussed above:

(a) Empowering managers with information technology by achieving the following:
- Develop an awareness program to help senior managers to become comfortable with using technology (technology adoption).
- Provide managers with "easy-to-use" information systems to help them to manage and make decisions (technology management).

(b) Helping people to cope with and to use technology in a way that ensures the following:

- The employees receive proper training to use technology and improve their technological performance.
- Adequate technical support is available for information systems.
- Advanced techniques to deliver education and training are used.
- Employee concerns about information technology are addressed through educating and providing guidelines to managers.

(c) Improving the performance of information management practitioners in order to:

- Improve the training of information management staff (e.g. Information management concepts, strategic planning, etc.).
- Promote sharing information on experiences.
- Establish information management as a career specialty.

11. Policy Implementation

The Information Manager is responsible for the implementation of this policy through appropriate management tools such as strategic planning, annual work plans and proposals. Furthermore, information management is the responsibility of every manager in their area as outlined in this policy review.

12. Summary of Information Management System for Mine Action in the Kurdistan Region

With the beginning of the mine action program in the Kurdistan region in 1999, the UNOPS established a comprehensive information management system. The system is designed in a way to allow managing of all mine related data. The initial step was to import all minefield information records from the MAG database. MAG has been involved in mine action in the region from 1992. Its major activity has been marking all minefields for future clearance. The new system was called UNOPS Mine Action Information System (UNOPS-MAIS).

The region's information database in the Kurdistan regions' mine action program was originally developed and designed to store level one survey information aimed at describing the overall mine and ERW problems in the region. Although the database met the initial requirements, it needed adaptation to further developments in the mine action sector. A number of requirements suggested that the database should be upgraded to fulfil such conditions. In order for the database to meet the requirements, it had to be upgraded to incorporate the current trends in mine action activities.

To upgrade the database to keep up with new trends and requirements a proper software engineering process was followed. After defining the functionality, an assessment procedure was conducted to determine which additional functionality needed to be implemented and which information aspects integrated to adhere to the new international mine action standards for humanitarian mine clearance operations. The assessment suggested the need to verify the quality of the existing data, correct or complete missing fields, modify the structure of the database to include new fields for blockage/impact data and to collect the corresponding data.

The new system involved the reporting of data from MAG to IKMAC and tasking command procedures from IKMAC to MAG. The migration of the MAG database to IKMAC-IMSMA enabled both parties to use the same database for planning and managing mine action operations. IKMAC and the National Mine Action Authority (NMAA) had all the additional MAG data available to provide better information to demining organizations. All information entered into this new system was automatically reflected in the Baghdad database each time database synchronization took place between Erbil and Baghdad. The full database with impact survey data was therefore available to plan for future operations. Care should be taken such that this confidential and detailed information is not made available to the Iraqi military for use in their future mine laying activities.

In conclusion, the database restructure process described above to verify and correct data, and collect additional data for the impact of landmines on communities through a follow-up survey, allowed the Kurdistan region mine action program to enhance its strategic management and its priority setting capability. In addition, it allowed for the implementation of new international mine action standards while continuing to providing service to the current information management needs of the program.

Notes

[1] See www.gichd.org, Accessed on Feb. 1, 2010.

[2] The flowchart is developed by the Swiss Federal Institute of Technology in Zurich as a part of IMSMA project development of Northern Iraq by UNSOPS. http://archiv.css.ethz.ch/services/imsma_EN.html.

[3] This section is based on a proposal for migrating the MAG database to the IMSMA database. The proposal was prepared by the co-author of this book as he was working as an information manager for UNOPS MAP and later for MAG in the Kurdistan region of Iraq.

[4] The mine action chain of events in the IMSMA. The information inside the red square is collected by the ILIS while the information inside the blue square is collected by demining organizations.

CHAPTER NINE

THE KURDISTAN REGION'S MINE ACTION SURVEY METHODOLOGY

The first LIS conducted in the region was based on the mine-affected communities in which their selection was determined based on the general survey results that in turn incorporated the data previously collected by the MAG. The general survey provided a set of mine-affected communities which served as a starting point for the impact survey. The purpose of conducting an impact survey was to gather blockage access data that was not originally captured during the general survey process.

The second LIS conducted by iMMAP (or VVAF) followed the same path as the UNOPS-LIS, and can be considered an updated process of this. Thus, in this chapter we explain in detail the methods and techniques used to conduct the UNOPS-LIS.

The process followed by the regional landmine impact survey was not a fully new LIS but rather a retro fit of an existing comprehensive minefields database with impact indicators. This has enabled the MAP to continue operations with minimal interference to utilize the interim results in current annual planning and tasking as well as to reconfirm the data that was collected during the previous general survey.

The impact survey employed the practices of systematic collection and analysis of "expert opinion" (Anjuman, Landowners, Mukhtar, etc.) to confirm the locations of communities impacted by landmines and ERW. The "impact interview," with its associated component activities of interviewing, mapping and assessment, was conducted in all mine-affected communities as determined by the general survey teams. The results of the community interviews were entered into the MAP database to form the basis for subsequent analysis.

The three governorates comprising the Kurdistan region create both a dominant Muslim religion and a tribal society. Care was taken to ensure that the survey teams did not offend local sensitivities or customs. The survey teams were recruited from the regions in the country where they would be expected to work and had previous experience gathering data for the general survey. Within one survey team, a female supervisor was included to facilitate comfort to female interviewees.

During the conduct of the landmine impact survey, several lessons were learned and are indicated later in this section resulting in refresher training for the survey teams, predominantly to reinforce training issues. The aim of this refresher training was to ensure data consistency throughout the survey process by virtue of improving interview/questioning techniques, improving ground assessments and improving size calculations/mapping skills. Management skills of the team leaders and location supervisors were also improved during the refresher training to ensure proper discharge of their respective levels of responsibilities.

1. The Impact Survey Teams and Site Selection

Based on the earlier work conducted by the general survey teams, information was available from the extensive mine action database indicating all governorates, districts, sub-districts and communities known to contain mines and ERW.

The MAP database was initially created in 1998 consisting of three tables and a series of lookup tables. Since this time, and until the use of the IMSMA, the database has evolved considerably to permit data entry for numerous mine action program forms (such as general survey) and the production of standard reports. Until summer 2001 two separate databases were managed within the program: a UNOPS Mine Action database and the Operations and Plans database (OPS & PLANS-DB). The merger of these databases was later migrated to the IMSMA. Further developments occurred in 2001 and 2002 including a database assessment by Cranfield University, the development of a prioritizing tool and the generation of a task order module.

From the 4,400 communities within the region a number of 1,096[1] communities were confirmed as mine-affected during the general survey and the remaining 3,304 communities were confirmed as mine-free. The confirmed mine-affected villages were selected for subsequent impact surveys, of which 1,022 were accessible for impact assessments, this number later increasing to 1,126 in the iMMAP-LIS stage.

Initially, a subset of 250 mine-affected communities (as determined from the general survey data) were selected based on analyzing nine factors, as indicated below, to serve as the starting point to collect the impact survey data:

- Number of victims over the last twelve months.

- Number of families to benefit directly from clearing a specific minefield within a specific village's boundary.
- Number of minefields currently present within a specific village boundary.
- Villages where ERW are presented and identified by the team.
- The total number of livestock involved in mines/ERW incidents within a specific village's boundary.
- The total number of families presently living in the village to benefit from clearing the minefields within that specific village's boundary.
- The total number of vehicles involved in mines/ERW incidents within a specific village's boundary.
- The intended land use as indicated per minefield within the specific village's boundary.
- Number of victims over the last five years.

These factors considered both community based and mined area characteristics. The factors were recognized as involving some duplication but enabling the first phase of the survey to proceed promptly, and be fed back into works planned for 2001. These initial villages served as interim weights until appropriate weights could be determined for the region with the local authorities, KRG and other UN agencies working throughout the area. To ensure consistent data quality throughout the landmine impact survey, these initial 250 communities were later re-surveyed.

The former general survey teams were trained early in February 2001 on completing the landmine impact survey. The survey officer conducted centralized training for twenty-four teams, and the teams were deployed on February 8, 2001 to conduct necessary fieldwork.

2. Community Interviews

Each impact survey team consisted of a team leader, a deputy and a driver. When the team arrived at a village they sought the village mayor or Anjuman, where they would explain the purpose of their visit. The Anjuman (or Mukhtar) would inform them of the landmine problem in the vicinity of the village, and efforts were made to establish contact with the landowner or victims. This approach was consistent with the strong location specific traditional and hierarchical culture present in the three northern governorates of Iraq.

Given the cultural norms, the males tended to be the most knowledgeable group predominantly working in the fields and assisted by family members during the harvest season, while the woman and children tended to focus on domestic household activities. Shepherds were difficult to interview given their roaming patterns. The village seniors with the greatest knowledge of landmines were the Anjuman. This formed a group of at least two informants to provide information collected in the impact survey form.

The interviews were conducted in the Kurdish language while the impact survey data was captured and registered on the prepared standard forms in English (see Appendix A). The data collection process went smoothly and no problems were encountered during the teams' interactions.

2.1 The Interview Process

A total of 2,133 key informants provided input to the overall data collection process. Table 9.1 below provides information about the informants who are classified by their category of occupations.

Table 9.1. Landmine impact survey interviewees

Villager Occupation	Number of informants interviewed
Villagers	811
Anjuman	497
Farmers	242
Head of Village	238
Mukhtar	171
Others	59
Shepherded	44
Military	26
Teachers	19
Workers	14
Religious Person	12
Total	**2,133**

A survey team typically interviewed, surveyed and mapped a minefield in two days or on average a whole community each week. Much of the time was spent interviewing the key informants. The remainder of the time in the local areas was spent conducting visual verification of individual mined areas from safe viewing points, and assessing or verifying the blockage factors. Additional time was spent within the community to gather other relevant victim incident data.

The group interview consisted of two parts: a confirmation of the mined site and a structured questionnaire interview. The surveyors completed the map sketch using resection techniques by taking bearings and distances to the corner points of the suspected mined area. The map helped the informants to communicate about the various mined areas in the community during the interview.

2.2 The Impact Survey Forms (Questionnaire)

The landmine impact survey form contained three components: (i) a village form, (ii) a minefield form, and (iii) an incident form. These are shown in detail in Appendix A and their contents are summarized below;

(1) Village form: This form covered the following information:

- Identification of village general information and population.
- Mine and ERW information.
- Killed livestock information.
- Destroyed vehicle information.
- External village support information.
- Additional information.
- Map sketch of route to village.

(2) Minefield form (per suspected mined area): Covered the following information:

- Survey team data.
- Geographic information.
- Minefield information.
- Road access data.
- Blocking access data.
- Additional comments.
- Map.

(3) Incident form (per case of incident): Covered the following information:

- Demographic information.
- Area information.
- Additional information.

3. The Field Editing Process and Data Entry

When the impact survey team leader completed the forms in the field, they were passed to the location supervisor to verify the accuracy of information. If errors or missing data was apparent, the forms were passed back to the impact survey team for completion. The location supervisors bundled the forms by village and passed the packages to the survey supervisor, who conducted random sampling validation. If acceptable, the forms were then passed to the data entry personnel for entry into the centralized database.

At the end of the week, all impact survey teams met with their location supervisors to review the completed surveys and assign new villages for assessment. These weekly meetings also provided an important group check on the work before the documentation was sent to the UNOPS survey supervisor and data entry to the centralized database.

After the documentation was received by the data entry section at the head office in Erbil, it was prepared by the data entry supervisor, and then passed to the six data entry clerks. The data was entered into a centralized database. The verification supervisor validated the data entry to ensure the information was correctly entered. If the data was entered accurately, the hard copies were stored at safe places for archival purposes.

4. The LIS Overview of the Kurdistan Region's Governorates

The results of the landmine impact survey indicated that the governorates of the Kurdistan region—Duhok, Erbil and Suleimaniya—suffer from many adverse consequences of landmines and ERW contamination. Although these impacts are felt more or less throughout these governorates, the frequent existence of dense "clusters" of high and moderate risk villages suggested that a well-planned and targeted set of mine action initiatives could make considerable improvements to the landmine situation in a relatively short period of time.

These sinister and gruesome remnants of conflicts have left the residents of the Kurdistan region's governorates with severe impacts to their daily lives. Based on the data collected during the landmine impact survey, the impacts were classified as high, medium and low are shown in Fig. 9.1.

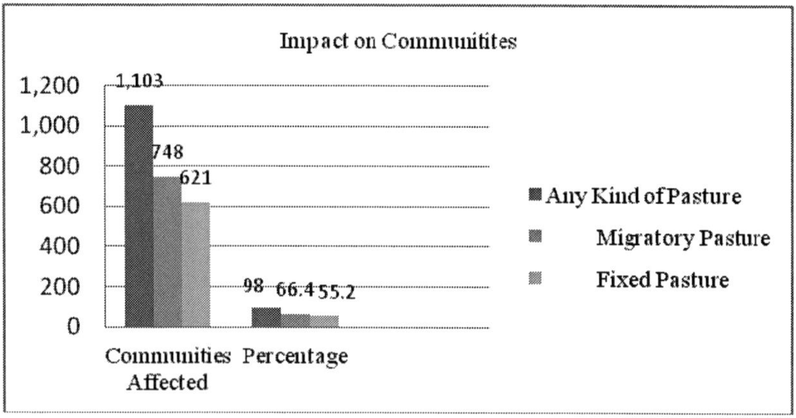

Fig. 9.1. High, medium and low impacts on communities in each governorate

The scope of this section is to analyze the impact of mines and ERW within the three governorates of the Kurdistan region. The analyses by governorate will consist of the following issues:

- Synopsis.
- Location.
- Scope and history of the landmine contamination.
- Impacts upon communities.

4.1 The Governorate of Duhok

Synopsis

The Duhok governorate is located in the north-west of the country, bordering Syria to the west, Turkey to the north, and Erbil to the east and south. The area of Duhok is approximately 9,920 square kilometres with a total population of approximately 727,927. Map 9.1 below shows the Mine/ERW contamination areas in the Duhok Governorate.

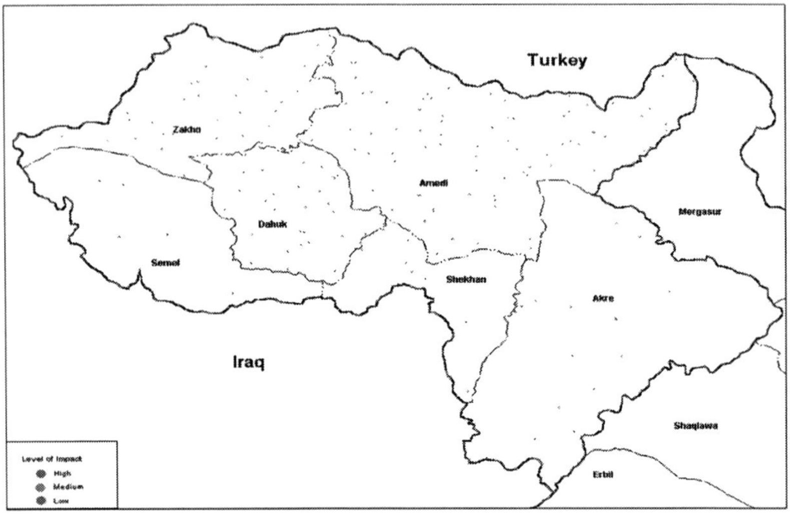

Map 9.1 Mine/ERW contamination area in Duhok Governorate

In the Duhok governorate, there are 178 mine-affected communities with a total mine-exposed population of 126,290. Seventeen of these communities are classified as having medium impact and 161 with low impact. A total of 563 mined areas with 27.4 million square metres have been identified. The majority of the land surface is either mountainous or flat land overgrown with short grass, bushes and/or trees. The landmine impact survey recorded a total of 241 mine/ERW victims in the communities, of which nineteen are from a recent period. Based on the landmine impact survey, 17% of the population in Duhok is affected by the presence of mines and ERW.

The Scope and the History of the Landmines Contamination

The landmine impact survey in the Duhok governorate has identified 178 mine-affected communities in six districts containing 563 suspected mined areas, blocking access to an estimated 27.4 million square metres of land. Levels of contamination vary from one district to another. The landmine-affected districtsvaries from Amedi with eighty four mine-affected communities, to Sumel with five mine-affected communities. The survey suggested that landmines and ERW in the governorate of Duhok originated during the 1991 Gulf War and were laid primarily along access routes to points of strategic importance and around former military positions. In some instances, local warring factions have also laid mines during their internal conflicts. Out of the 27.4 square km mined area in the Duhok Governorate, 23 square kilometres are mined with AP landmines. There are no communities containing only AV mines, however mixed AV and AP mines account for 4 square km. Table 9.2 below provides more detailed information on AP and AV mines in the Duhok governorate.

Table 9.2. AP and AV presence in the Duhok governorate

Type of Device	No of Communities	Mined Areas	Area in sq m
Anti-personal (AP)	168	528	23,330,414
Anti-vehicle (AV)	0	0	0
Mixed	8	30	3,993,800
Unknown	2	5	110,000
Total	178	563	27,434,214
UXO & Mines	48	101	4,713,720

Mine and ERW Victims

The survey found that 241 individuals in the governorate of Duhok were injured or killed by mines, 19 in the years 1999 and 2000 and remaining 222 persons during previous periods. It is interesting to note that 18 out of the 45 victims were fighting in a war, the reason for this possibly being ongoing internal conflicts. Furthermore, mine accidents related to livestock correlates with the above observation where grazing land is denied. Ten of the abovementioned mine victims were injured when tampering with mines and ERW. Table 9.3 below shows the distribution of activities of the victims at the time of accident.

Table 9.3. Sub-set of major activities at time of accident

Activity at time of the accident	Male	Female	Total
Fighting in a war	18	0	18
Grazing animals	9	0	9
Tampering with Mine/UXO	9	1	10
Walking	7	1	8
Total	**43**	**2**	**45**

Mine Risk Education

The mine risk education program delivers safety lessons and organizes mine risk information sessions for the local population in the Duhok governorate. It is expected that the program will play a significant role in reducing the landmines/ERW accidents. Since its inception in December 2000, MAP-MRE made the following achievements reported in Table 9.4 below, limited to the Duhok governorate.

Table 9.4. MRE activities in Duhok governorate

Training Locale	No. of visits	Beneficiaries			Total
		Men	Women	Children	
IDP Camp	10	891	846	2,375	4,112
Collectives	13	1,300	999	2,319	4,618
Villages	220	7,702	8,639	14,265	30,606
Total	243	9,893	10,484	18,959	39,336

Impact on Communities

The contamination of landmines/ERW affects 178 communities with a total population of 126,290. The settlement types affected are villages, collective towns, and sub-district centres. Seventeen communities are considered medium impact, 161 communities suffer low levels of impact, while no community has been identified with high impact during the survey process. Fig. 9.2 below provides the number of communities (upper position), victims (middle) and minefields (circle size) and estimated areas contaminated (lowest) in the Duhok governorate.

The medium and low impact communities differ by their characteristics in form of victims, contamination area size, and population size. In the case of medium impact communities, the landmine/ERW impact survey identified 17 communities in four districts with a total population of 29,545. In the cases of low impact communities, the landmines/ERW impact survey indicated 161 low impact communities throughout the six districts with a total population of 96,745.

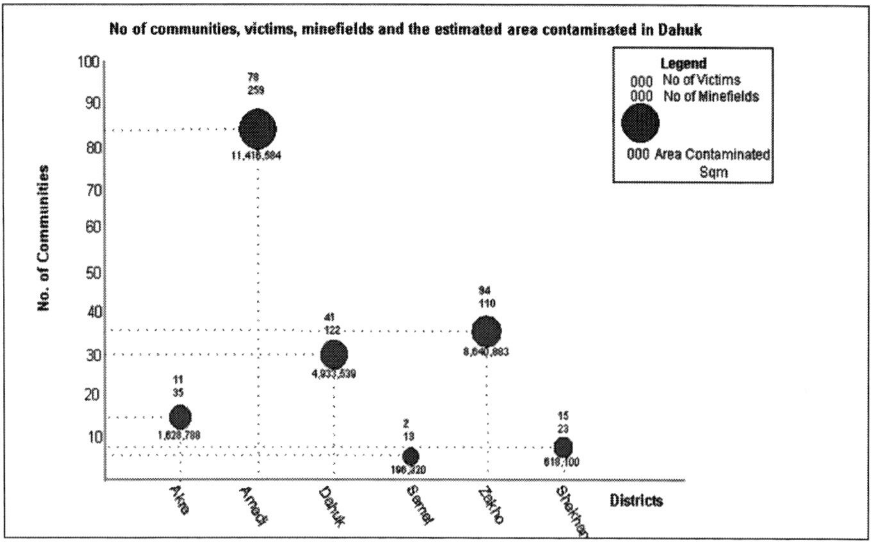

Fig. 9.2 Number of Mine/ERW Victims, Number of Mined Areas and Estimated Area Contaminated in Duhok Governorate

From the available statistics, it is observed that there is no high correlation between population concentration and landmine contamination areas. Fortunately, the centre of intensity of conflicts was not always located around districts with a high population concentration. The statistics about the number of affected communities, victims, and estimated area contaminated in Duhok governorate distributed across different districts within the governorate is shown in Table 9.5 below.

Table 9.5. Number of affected communities, victims and estimated area contaminated in Duhok governorate

District Name	No. of Communities				No. Recent Victims	Total No. of Victims	No. of Mine fields	Area Impact Surveyed m^2
	High	Med.	Low	Total				
Akre	0	0	15	15	3	11	35	1,628,788
Amedi	0	6	78	84	10	78	259	11,416,584
Dahuk	0	1	29	30	0	41	122	4,933,539
Semel	0	0	5	5	0	2	13	196,320
Shekhan	0	1	8	9	1	15	23	618,100
Zakho	0	9	26	35	5	94	110	8,640,883
Total	0	17	161	178	19	241	562	27,434,214

In the majority of affected communities, landmine/ERW contamination blocks access to grazing and crop farms. As many as 17 out of the 178 communities reported blocked access to grazing, while another 39 reported blocked access to crop farms. Among the other types of blockages, seven communities reported blockages to hunting land, shelter material and wild plants, and another nine reported blockage to water sources. The number and percentages of communities by type of impact in the Duhok governorate is reported in Table 9.6 below.

Table 9.6. Number and percentage of communities by type of impact in Duhok governorate

Group	Type of Impact	No. of Affected Communities	Percentage of Affected Communities
A	Crop	39	37.5
B	Grazing	17	16.3
C	Water Resources	11	10.6
D	Firewood Collections	10	9.6
E	Infrastructure	27	26.0
	All types	104	100.0

The presence of mines and ERW has affected different aspects of the local population in the Duhok governorate, varying from crops to grazing, water resources, etc. The number of communities affected by the type of impacts in the above table is different than the total number of communities affected by mines, since multiple occurrences may occur in the database. These are illustrated in Fig. 9.3 below.

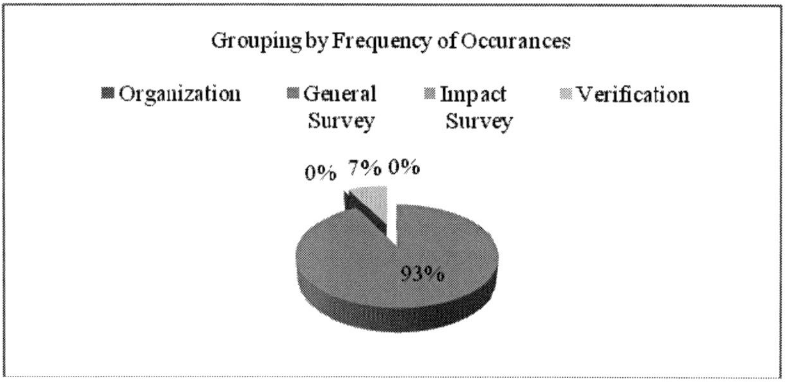

Fig. 9.3. Frequency of occurrence by blockage factors in Duhok governorate

4.2 The Governorate of Erbil

Synopsis

In the Erbil governorate, there are 299 mine-affected communities with a total population of 266,027. The landmine impact survey indicated that mines affect 18% of the total population in the Erbil governorate. Map 9.2 shows the mine/ERW contamination areas in the governorate.

Map 9.2 Mines and ERW Contamination in Erbil Governorate

A total of 784 mined areas with a surface area of 331 million square metres have been identified in Erbil governorate. The majority of the affected areas in the Erbil governorate are either hilly or mountainous and overgrown with grass, bushes and/or trees. The landmine impact survey recorded a total of 885 mines/ERW victims in the affected communities of which 81 are from more recent times. The affected communities are classified as 7 with high impact, 37 with medium impact, and remaining 219 with low impact. The distribution of number of communities with different levels of impacts is shown in Fig. 9.4 below.

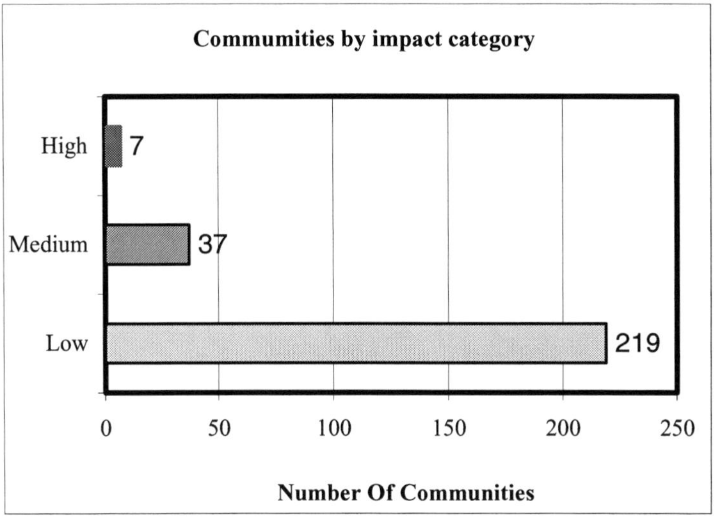

Fig. 9.4. Number of Communities with High, Medium and Low Impact in Erbil Governorate

Geographical Location

The Erbil governorate is located in the north-east of the country, bordering Iran and the governorate of Duhok to the north and north-west, the Governorate Suleimaniya to the south and south-east and the governorate of Kirkuk to the south. The governorate of Erbil spans a total area of 10,620 square kilometres containing a population of 1,155,883 people. Erbil is the capital city of the Kurdistan region and its population is growing fast. In addition to the regional governance infrastructure, it hosts several Western countries' consulates and offices of a large number of international corporations and many non-governmental organizations operating in the region. Concerning this high density of productive infrastructures, the landmine contamination can thus have a devastating effect on the region's economy.

The Scope and History of the Landmines' Contamination

The landmines in the governorate of Erbil were laid primarily along access routes to points of strategic importance during the 1980–1988 Iraq-

Iran conflict and the subsequent internal conflict between the Kurdish local political factions. The landmine impact survey identified 263 mine-affected communities containing 784 mined areas blocking access to an estimated 108.8 million square metres of land in six districts of the Erbil governorate. Mkhmur is the only district in the governorate that is not affected by the presence of mines/ERW. The Mkhmur district is presently under the control of the Iraqi government forces which may explain the discriminative policy in the mine laying policy. It indicates that mine laying was a measure of ethnic cleansing targeting the Kurdish population rather than defence related. Levels of contamination vary from one district to another. The landmine-affected districts vary from the Soran district with 94 affected communities to the Erbil district with 7.

Nearly 105 square km of the mined area in the governorate of Erbil is reported to contain anti-personnel landmines. Mixed areas containing AP and AV mines account for 3.5 square km. Table 9.7 below shows the distribution of communities, mined areas, and their areal size in the Erbil governorate.

Table 9.7. AP and AV presence in the Erbil governorate

Type of Device	No of Communities	Mined areas	Area in m^2
AP	249	734	105,151,447
AV	5	16	874,312
AV and AP	9	34	2,812,811
Total	263	784	108,838,570
UXO	71	274	42,119,225

Mine and ERW Victims

The survey found that 885 victims in the governorate of Erbil were injured or killed by mines, of which 81 occurred in the years 1999, with the remaining 804 prior to this. Fifty-two victims were injured while tampering with mines/ERW, sixty-one while walking and mines/ERW that self-exploded have injured another twenty-four. The distribution of activities of victims at time of an accident by gender is reported in Table 9.8. As mentioned previously, due to the traditional division of labour in the region, the majority of the victims are male.

Table 9.8. Sub-set of major activities at time of accident

Activity at time of the accident	Male	Female	Total
Walking	53	8	61
Grazing animals	21	0	21
Tampering with Mine/UXO	49	3	52
Nothing, it just exploded	24	0	24
Total	147	11	158

Mine Risk Education

Similar to the situation in the Duhok governorate case, the mine risk education program delivers safety lessons and organizes mine risk information sessions for the local population in the Erbil governorate. The program is expected to play a significant role in reducing the frequency of landmines/ERW accidents. However, the measure of the final outcome of such preventive measures was not reported. Since its inception in December 2000, MAP-MRE has made the following achievements in the Erbil governorate (see Table 9.9 below).

Table 9.9. MRE activities in the Erbil governorate

Training Locals	No. of visits	Beneficiaries			Total
		Men	Women	Children	
IDP Camp	2	5	5	544	554
Collectives	14	723	750	13,302	14,775
Villages	138	9,308	10,039	20,909	40,256
Total	**154**	**10,036**	**10,794**	**34,755**	**55,585**

Impacts on Communities

Landmine/ERW contamination affects 263 communities in the Erbil governorate with a total population of 207,490. The most affected settlement types are villages, district centres and sub-district centres. The impacted communities are classified as high, medium, and low impact in accordance with the landmine impact survey scoring mechanism.

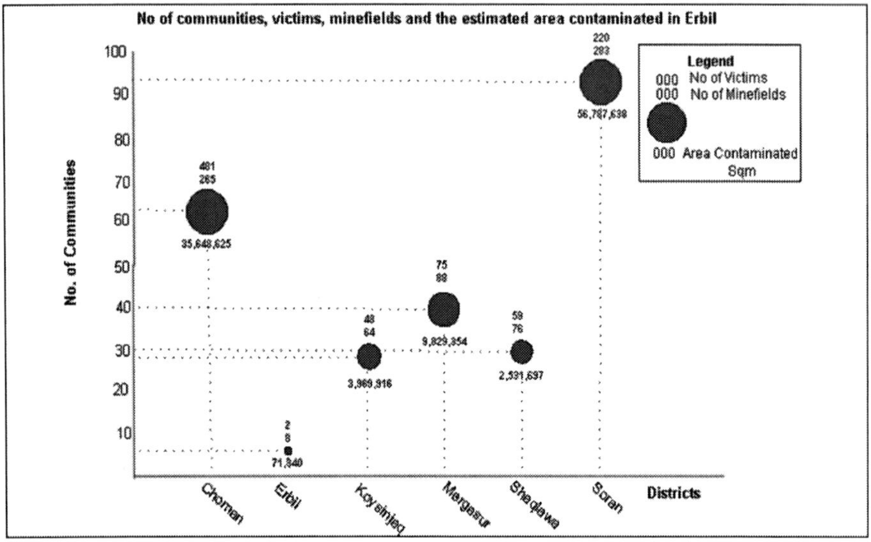

Fig. 9.5 Number of Mine/ERW Victims, Number of Mined Areas and Estimated Area Contaminated in Erbil Governorate

Fig. 9.5 above shows the number of mine/ERW victims, the number of mined areas and the estimated area contaminated in the Erbil governorate at the district level. The communities in the Erbil governorate subjected to different levels of impact and their characteristics are described below.

High impact communities: The landmine/ERW Impact survey identified seven communities in the Choman district and one community in the Soran district with a population of 49,510 with high impact.

Medium impact communities: The landmine/ERW impact survey identified thirty-seven communities in the Choman, Koysinjaq, Shaqlawa and Soran districts with a total population of 68,555 with medium impact.

Low impact communities: the landmine/ERW impact survey indicated a number of 219 communities in the Choman, Margasur, Mkhmur and Soran districts with a population of 89,425 with low impact.

Distribution of the number of affected communities, victims and areas of impact in the Erbil governorate is reported in Table 9.10 below. The number of communities is further divided into different groups by level of impact. The Chonan and Soran districts are the two main impacted districts in the Erbil governorate.

Table 9.10. Number of affected communities, victims and areas of impact in Erbil governorate

District name	No. Communities				No. Recent Victims	Total No. Victims	No. Mine fields	Area Impact Surveyed m^2
	High	Med.	Low	Total				
Choman	6	18	40	64	57	481	265	35,648,625
Erbil	0	0	7	7	0	2	8	71,340
Koysinjag	0	5	23	28	3	48	64	3,969,916
Mergasur	0	1	38	39	2	75	88	9,829,354
Shaqlawa	0	3	28	31	3	59	76	2,531,697
Soran	1	10	83	94	16	220	283	56,787,638
Total	7	37	219	263	81	885	784	108,838,570

Estimated Area Contaminated in the Erbil Governorate

In the majority of affected communities, landmine/ERW contamination blocks access to grazing land, firewood collection area and wild plants. As many as 81 communities reported blocked access to grazing land while 124 reported blocked access to firewood collection areas, wild plants and hunting lands (field collection), while 31 communities reported blockages to crop lands, and 24 to water resources. Blockage of water resources is probably the most dangerous and worst type of landmine impact on the surrounding communities' well-being. Laying mines near springs and water resources were found to be the most effective measure of depopulating the region through forced migration. The type of impact is reported in Table 9.11 below.

Table 9.11. Number and percentage of communities by type of impact in the Erbil governorate.

Group	Type of Impact	No. of Affected Communities	Percentage of Affected Communities
A	Crop	31	11.1
B	Grazing	81	29.0
C	Water Resources	32	11.5
D	Firewood Collections	124	44.4
E	Infrastructure	11	3.9
	Total	279	100.0

The presence of mines and ERW affected different aspects of living conditions of the local population in the Erbil governorate in varying forms, from crop farms to grazing land, water resources, etc. The total affected communities in the above chart indicate that one community could be affected by more than one type of impact. These are illustrated further in Fig. 9.6 below.

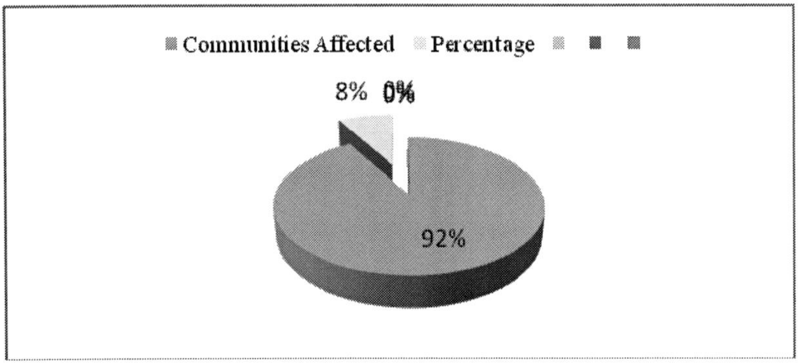

Fig. 9.6. Grouping the type of impact by Frequency of Occurrences in Erbil Governorate

4.3 The Governorate of Suleimaniya

Synopsis

In the Suleimaniya governorate there are 582 mine-affected communities with a total 409,530 people living within the area. These communities were classified as 28 with high impact, 145 with medium impact and remaining 409 with low impact. A total of 2,096 mined areas with a surface of 325.9 million square metres were identified. The majority of the reported surface is either mountainous, such as the Sara mountain chain, or flat land but overgrown (at least in parts) with short grass, bushes and/or trees. The survey also recorded a total of 2,323 mine/ERW victims in affected communities, of which 215 been subjected to mine accidents in recent years. Map 9.3 shows the Mine/ERW contamination in the Suleimaniya governorate.

Geographical Location

The Suleimaniya governorate is located in the north-east of the country bordering Iran to the east, the governorate of Erbil to the north and west and the governorate of Kirkuk to the south. The governorate of Suleimaniya is the largest governorate in the region spanning an area of 20,680 square kilometres and containing a population of approximately 1,663,190 people.

The Scope and History of the Landmine Contamination

The landmine impact survey suggested that landmine and ERW in the governorate of Suleimaniya originated primarily during the 1980–1988 Iraq-Iran War and the subsequent local political factions' internal conflict. Landmines were laid primarily along access routes to points of strategic importance and around the concentration of the former military positions within the region. The survey reported a total of 581 affected communities in the Suleimaniya governorate containing 1,902 mined areas. These mined areas are reportedly blocking access to an estimated 203.3 million square metres of land.

All of the twelve major districts in the Suleimaniya governorate are more or less affected by landmines/ERW, and levels of contamination differ from one district to another. The landmine-affected districts vary from Sharbazher with 164 affected communities to Kiffri with only eight. As discussed before, among the three governorates the frequency and type of accident is highest in the Suleimaniya governorate, compared to the Duhok and Erbil governorates. It might be due to the location of the governorate in conjunction with bordering Iraq-Iran, and thus it could be expected that the governorate is more frequently subjected to battlefields in the internal and external conflicts.

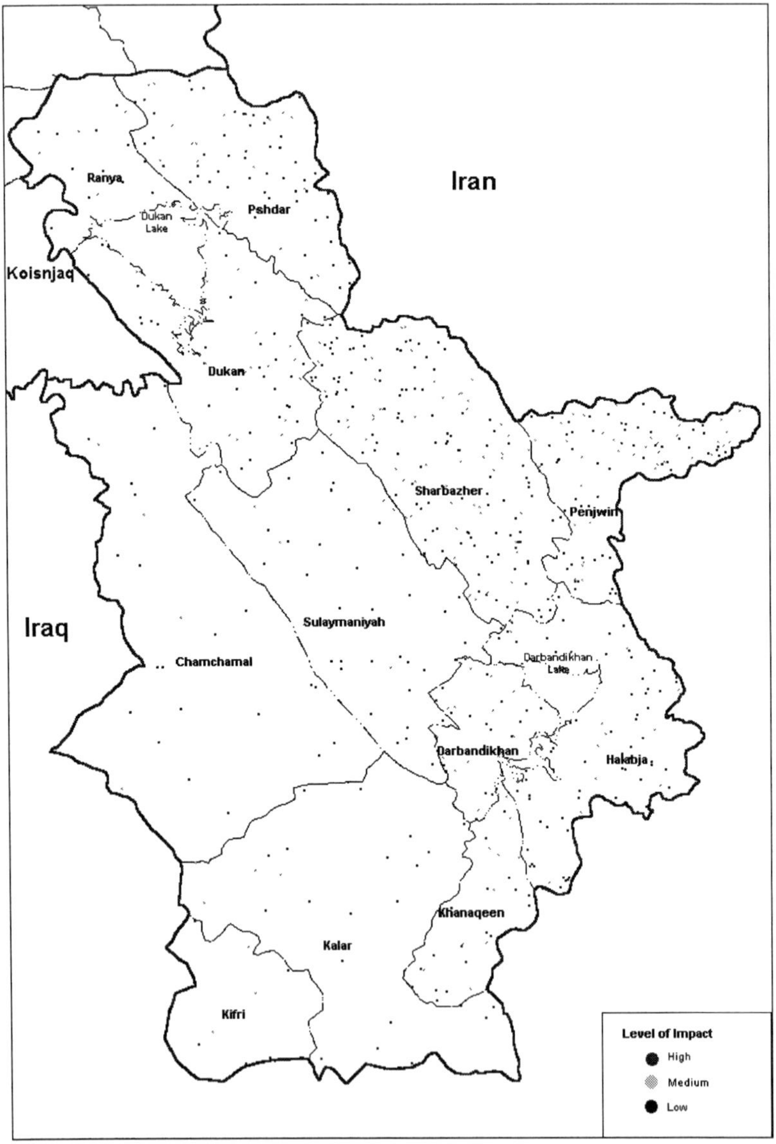

Map 9.3 Mine/ERW contamination in Suleimaniya Governorate

The communities in the Suleimaniya governorate are classified by the levels of impact. Fig. 9.7 below shows the frequencies of accident by type.

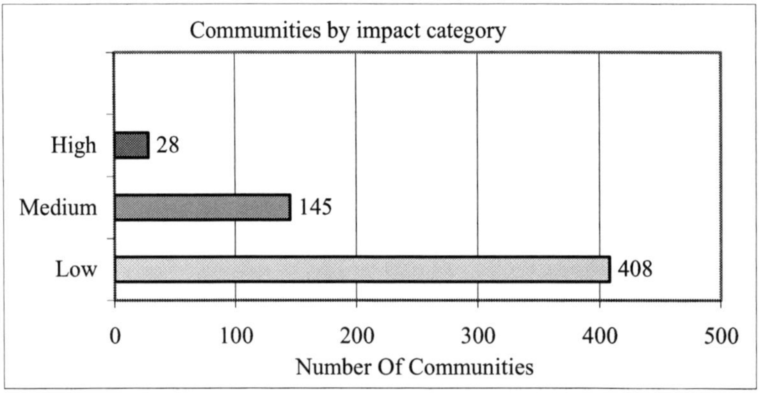

Fig. 9.7. Communities with high, medium and low impact in the Suleimaniya governorate

About 175 square km area in the governorate of Suleimaniya are mined with AP. The mixed minefields containing AP and AV mines account for nearly 27 square km. Table 9.12 below reports the distribution number of impacted communities, mine areas, and their sizes distribution by type of mine. The statistics show that AP mines are the dominating type followed by mixed AP and AV devices. The mountainous character of the governorate is the factor that leads to the frequent use of anti-personnel type of devices to reduce mobility of the population and preferably impose forced migration on the governorates with a dominant Kurdish population.

Table 9.12. AP and AV presence in the Suleimaniya governorate

Type of Device	No of Communities	Mined areas	Area in m^2
AP	541	1721	174,473,749
AV	8	14	1,902,475
Mixed	30	162	26,831,620
Unknown	2	5	111,957
Total	581	1,902	203,319,801
UXO & Mines	431	1,267	137,530,696

Mine and ERW victims

The survey found that 2,323 victims in the governorate of Suleimaniya were injured or killed by mines, with 215 in the years 1999 and 2000 and remaining 2,108 prior to this. Out of the total mine victims, 163 were injured when tampering with a mine or ERW, while the other notable groups are victims due to grazing and walking activities at the time of accident. The number of accidents distinguished by type of activity and gender in the governorate is reported in Table 9.13 below.

Table 9.13. Sub-set of major activities at time of accident

Activity at time of the accident	Male	Female	Total
Walking	47	5	52
Grazing animals	56	0	56
Tampering with Mine/UXO	155	8	163
Nothing, it just exploded	16	7	23
Total	274	20	294

Mine Risk Education

The mine risk education program delivers safety lessons and organizes information sessions for the local population in the Suleimaniya governorate that are subjected to such risk. The program is expected to play a significant role in reducing landmines/ERW accidents. In later chapters we will deal with assessment impacts of programs. Since its inception in December 2000, MAP-MRE has made the following achievements in the Suleimaniya governorate (see Table 9.14 below).

Table 9.14. MRE Activities in Suleimaniya Governorate

Training Locals	No. of visits	Beneficiaries			Total
		Men	Women	Children	
Villages	204	8,393	7,994	12,861	29,248
Total	204	8,393	7,994	12,861	29,248

Impact on Communities

The landmine/ERW contamination affects 581 communities in the Suleimaniya governorate with a total population of 409,530. The affected settlement types consist of villages, collective towns, sub-districts and district centre settlements. The presence of mines and ERW in the

Suleimaniya governorate has left high, medium, and low impacts on communities. The distribution of impacted victims, minefields, contaminated areas and their sizes is reported in Fig. 9.8 below.

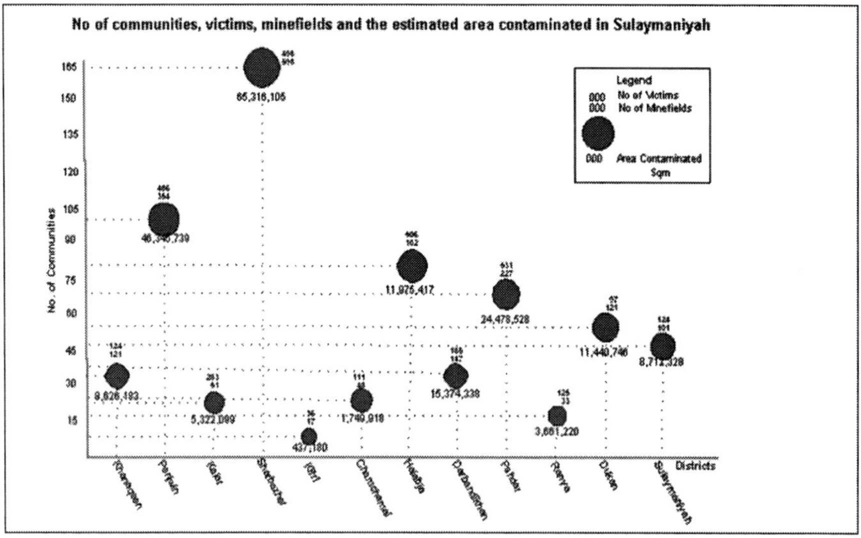

Fig. 9.8 Number of Mine/ERW Victims, Minefields and Estimated Size of Contaminated Area in Suleimaniya Governorate

The communities in the Suleimaniya governorate, grouped by different levels of impact and their characteristics, are described below:

High impact communities: The landmines/ERW impact survey identified 28 communities in eight districts with a population of 105,110 families who suffer a high level of impact in the governorate of Suleimaniya.

Medium impact communities: The landmine/ERW impact survey has reported 145 communities in twelve districts with a population of 145,435 with a medium level of impact.

Low impact communities: The landmine/ERW impact survey reported 408 communities in twelve districts with a population of 158,985 with a low level of impact.

The distribution number of affected communities, victims and estimated area contaminated in the Suleimaniya governorate further divided by the type of impact and districts is reported in Table 9.15 below.

Table 9.15. Number of affected communities, victims and estimated area contaminated in Suleimaniya governorate

District name	No. Communities				No. of Recent Victims	Total No. of Victims	No. of Mine fields	Areas of Impact surveyed m^2
	High	Med.	Low	Total				
Chamcahmal	1	2	22	25	6	71	48	1,749,918
Darbandikhan	2	13	20	35	2	146	147	15,374,338
Kalar	1	3	19	23	5	161	51	5,322,099
Khanaqeen	0	9	24	33	5	104	102	8,626,183
Kiffri	1	2	5	8	13	30	17	437,180
Dukan	0	15	41	56	15	69	121	11,440,746
Halabja	2	11	10	23	12	246	102	11,975,417
Penjwin	3	28	66	97	23	465	354	46,225,739
Pshdar	7	9	38	54	36	427	227	24,478,528
Ranya	0	2	14	16	5	55	34	3,661,220
Sharbazher	11	46	108	165	82	502	598	65,316,105
Suleimaniya	0	5	41	46	11	47	101	8,712,328
Total	28	145	408	581	215	2,323	1,902	203,319,801

In the majority of mine-affected communities, mine/ERW contamination blocks access to grazing land and cropland. Table 9.16 below shows that blocked grazing land affected 218 communities, while 192 reported blocked cropland. Another 250 communities reported blockage to firewood collections and 94 communities to water resources.

Table 9.16. Number and percentage of communities by type of impact

Group	Type of Impact	No. of Affected Communities	Percentage of Affected Communities
A	Crop	192	23.4
B	Grazing	218	26.6
C	Water Resources	94	11.4
D	Firewood Collections	250	30.4
E	Infrastructure	67	8.2
	Total	821	100.0

The presence of mines and ERW has affected different aspects of the local population in the Suleimaniya governorate, varying from crops to grazing and water resources as illustrated in Fig. 9.9 below.

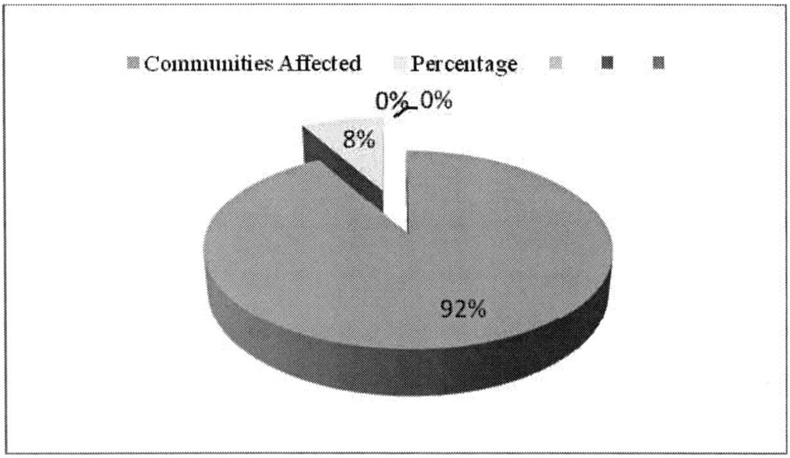

Fig. 9.9. Frequency of occurrence by blockage factors in the Suleimaniya governorate

5. Summary of the Kurdistan Regions Mine Action Survey Methodology

This chapter introduced the landmine impact survey conducted in the Kurdistan region based on its mine-affected communities. The mine action survey, mine action program, composition of mine action groups and their work procedure and data collection and management was described in detail. The impact survey employed the practices of systematic collection and analysis of data in cooperation with the community's representative to confirm the locations of communities impacted by landmines and explosive remnants of war. The impact interview, mapping and assessment were conducted in all mine-affected communities determined by the general survey teams. The results of the community interviews were entered into a database, which formed the basis for subsequent analysis.

The data collection and activity results are presented in this chapter for the entire region. During the conduct of the landmine impact survey, several lessons were learned which resulted in additional training for the survey teams, predominantly to reinforce training issues. The aim of this training was to ensure data consistency throughout the survey process by virtue of improving interview/questioning techniques, ground assessments and size calculations/mapping skills. Management skills of the team leaders and location supervisors were also improved during the training provided to ensure proper discharge of their respective levels of responsibilities.

Based on earlier work conducted by the general survey teams, information was available from the extensive mine action database covering all governorates and distributed into districts, sub-districts, and communities known to contain mines. The MAP database was initially created in 1998 and evolved considerably to permit data entry for numerous mine action program forms and the production of standard reports. Of the 4,400 communities within the region, 1,096 communities were confirmed as being mine-affected. The confirmed mine-affected villages were selected for subsequent impact surveys, of which 1,022 were accessible for impact assessments, later increasing to 1,126. Initially, a subset of 250 mine-affected communities was selected based on analyzing nine factors. These factors consider both community based and mined area characteristics.

A total of 2,133 key informants provided input to the data collection process. A survey team typically interviewed, surveyed and mapped a minefield in two days or on average a whole community each week. Much of the time was spent interviewing the key informants. The remainder of the time in the locality was spent conducting visual verification of individual mined areas, and assessing or verifying the blockage factors. The group interview consisted of two parts: a confirmation of the mined site and a structured questionnaire interview. The landmine impact survey form contained three components, a village form, a minefield form (per each individual field), and an incident form (per each individual incident).

When the impact survey's team leader completed the forms in the field, they were passed on to the location supervisor to verify the accuracy of information. After the documentation was received by data entry section, it was prepared by the data entry supervisor, and then passed to data entry clerks. The data was then entered into a centralized database and verified.

The result of the landmine impact survey is overviewed in a disaggregated form. The region is divided into its three constituent governorates further divided into different districts and by the magnitude of impact or frequency of accidents. The accidents are classified by the outcome, characteristics of the victims, and the type of activities at the time of accident. The result indicates that the three governorates of Duhok, Erbil, and Suleimaniya suffer from many adverse consequences of landmines and ERW contamination. Although these impacts are felt more or less throughout these governorates, and the frequent existence of dense clusters of high and moderate risk villages suggests that a well-planned and targeted set of mine action initiatives could make considerable improvements in the landmine situation in a relatively short period of time and at low cost. It will have a strong impact on safety and quality of life of citizens in the region.

A final recommendation is that the regional landmine agency should be cautious in the distribution of the collected data to the central government landmine agency. The information provides detailed information about locations of landmines and their impacts which can have devastating effects if such sensitive information is accessed by the army and security forces of Iraq. There is no doubt that these forces will use the information in possible future mine laying activities against the Kurdish population.

Notes

[1] All the figures indicated in this chapter refer to the data addressed by UNOPS DB before the second LIS was conducted.

CHAPTER TEN

STATISTICAL ANALYSIS OF LANDMINE FATALITIES IN THE KURDISTAN REGION

This chapter analyzes mines and unexploded ordnance (UXO) victim data in the Kurdistan region of Iraq collected from the 1960s to 2005. In addition to descriptive analysis of the data, we use regression analysis to identify the determinants of the probability of getting killed by a mine or UXO and to estimate the determinants' impacts. The rates of killed and injured victims are explained using a set of socio-economic variables. Since the data is repeated in cross sections in which individuals are observed when they are subjected to accident and in order to account for the dynamic aspect of the process and heterogeneity by location, we create pseudo panel data where districts are observed over the entire study period forming almost a balanced panel data. This allows for analysis of both heterogeneity and dynamics across socio-economic characteristics, locations and over time, as well as to control for unobserved location and time effects. The results show that women and children are exposed to a high level of mine risk. It can also be useful in planning, monitoring, and resource allocation for mine action and labour market programmes and rehabilitation activities.

Keywords: Landmine, unexploded ordinance, pseudo panel data, security and war; health, Kurdistan;
JEL Classification Numbers: C23; C42; H56; J15; N45;

1. Introduction

The explosion of all kinds of mines and UXOs will cause death or injury to humans and animals. In addition, it causes defects such as amputation and paralysis, which in turn lead to strong psychological, social, and economic effects. These effects are on the individual, family, and society and will turn people with disability and impairment into persons unable to work, walk or move. It is well known that, despite the occurrence of daily incidents in many places around the world, the full impact will start after the return of normal life to the contaminated rural areas. Thus, the real impact is expected to take place following the return of the people to their villages and resuming their normal daily life activities such as agriculture and animal husbandry, where shepherds and peasants suffer the most. Another area in which people are subjected to danger is when the maintenance and rehabilitation of damaged infrastructure starts in the remote areas that are affected by massive deployment of mines and UXOs.

The Kurdistan region is one of the major mine and UXO densely contaminated areas. The number killed per 1,000 population varies in the interval 0.27 and 58.83 among the different districts. The number and density of mines and UXOs in the region is not less if we compare it with other heavily mine-affected countries. Cambodia for example, reported 600 victims per month, and the proportion of people with disabilities is 1 to 231 of the population. The corresponding number in Afghanistan is 500 victims in a week. Statistics show that there are more than 75,000 victims in Angola that are disabled by mines and UXOs.[1] It is believed that laying landmines in Iraq dated back to the 1940s, but it was sporadic and did not appear to influence civilian life comprehensively. However, its real beginning was from conflicts between the Iraqi army led by Saddam Hussein's regime and the Kurdish liberation forces (Peshmarga). Although the number of landmines was limited, it had serious consequences on citizens. The process of laying landmines continued by the Iraqi regime on the roads and axes, in the agriculture lands, around spring water sources, and other places aimed to limit the movement of the Peshmarga.

In the Kurdistan Region of Iraq, with a population of nearly five million, mines were laid during four distinct phases. The first phase started in the mid-1970s to subdue Kurdish resistance movements operating in rural areas. The second phase was during the Iraq-Iran war in 1980s, when the number of landmines increased dramatically along the border with Iran. The third phase was the Gulf War when mines were laid along the border with Turkey. The fourth phase was related to the years from the end of the Gulf War, when Kurdish factions used landmines against each other.[2] Thus, the Kuwait-related Gulf War, the invasion of Iraq by allied forces, and the Iraqi government's hostile assimilation policy resulted in laying a huge number of landmines across the border, especially within the Kurdish territories in the north. The Iraqi army has knowledge about the mines and the risks associated with them and it has access to equipment to clear the minefields. In general, the armed forces laid mines to depopulate specific regions. Therefore, they are reluctant to supply minefield maps to mine-clearing agencies or to participate in the clearing activities.

This study attempts to analyze mines and UXO victim data in the Kurdistan region of Iraq during the period of the 1960s to 2005. In addition to descriptive analysis of the data, we use panel data regression analysis to identify the determinants of the probability of death from mines/UXOs and to estimate the determinants' impacts. Variations in the rates of killed and injured victims are explained using a set of socio-economic variables. Since the data is repeated in cross sections in which individuals are observed when they are subjected to accident and in order to account for the dynamic aspect of the process and heterogeneity by regional location, we create pseudo panel data where districts are observed over the entire study period. This allows for the analysis of both heterogeneity and dynamics across locations and over time, as well as for the control of unobserved location and time effects. The result can be useful to regional governments and NGOs in planning, monitoring and resource allocation for mine action and rehabilitation activities.

The rest of the chapter is organized as follows. A brief review of the literature is provided in Section 2. In Section 3 and 4 we describe the repeated cross-sectional data and its transformed pseudo panel data. The model is outlined in Section 5, and the result analyzed in Section 6. In Section 7 we discuss the socio-economic, psychological, and overall impacts on health and well-being of the mine/UXO victims. The final section summarizes the study.

2. Review of the Literature on Landmines

To the best of our knowledge, little academic research has been published regarding factors determining the occurrence of landmine deaths and injuries or analysis of landmine accident risk factors at the community level (see also Krug et al. 1998). According to Ascherio et al. (1995), evidence based on hospital and household surveys shows that the return of civilians to landmine-contaminated areas, their contact with the mines, and their engagement in agriculture will increase the possibility of landmine accidents, in addition to demining activities during mine clearance. Regarding the activity of the mine victims at the time of the accidents, Roberts and Williams (1995) concluded that most of the victims were injured while walking in mine-suspected areas. Shabila et al. (2010) analyzed the characteristics of hospitalized landmine injured patients in the Erbil governorate of Kurdistan. They found the need to examine the reasons behind handling explosives. Jahanlu et al. (2002), through investigating mortality in landmine accidents in Iran, found that most fatalities seemingly occurred in the pre-hospital setting.

Other research investigated the gender aspect of landmine accidents. Berthiaume (2003) explained that men and women have different mobility patterns, and hence have different exposure and vulnerability to mines and UXOs. Women will most likely be exposed to landmine accidents while gathering wood and fuel or water, while men are exposed while farming and travelling on public roads. A report by Bottomly (2003) showed that 61% of victims in Cambodia are men and boys who are mostly exposed to mine risk. The pattern is similar in different parts of the world where elements of age and gender are observed in most surveys related to landmines. In a recent study, Bendinelli (2009) compared the impact of explosive remnants in rural Cambodia. Results show that 26.4% of the victims were children, predominately female, who sustained the more severe injuries.

Quesada (1998) reported on how the ripple effects of war and its aftermath are felt even after being mediated by time, space and social status. Through a case study of a Nicaraguan boy and his family, Quesada argued that the legacy of war, structural violence and endemic poverty are chronic. The impact of trauma and trauma care parameters on chronic pain in landmine accident survivors in Cambodia and Kurdistan shows high levels of chronic pain more than one year after the accident. Injury severity and primary trauma care effect chronic pain, and patient loss of income is correlated with chronic pain syndromes (Husum et al. 2002). Champion et al. (1989) presented revised versions of trauma scores for use in outcome predictions and evaluations of patients with serious head injuries.

Analysis of the direct and indirect consequences of landmines on public health shows that landmines constitute a major public health problem in the world (Kakar et al. 1996). A number of measures and strategies are presented to prevent accidents and to increase the effect of treatment and rehabilitation programs. The environmental health consequences of anti-personnel landmines globally were reviewed by Newman and Mercer (2000). They recommended political efforts to ban the use of landmines. Husum et al. (2002) reported that the economic standing of the victim's family deteriorated after the accident. The social cost of landmines differs across countries suffering from landmines. Anderson et al. (1995) showed that landmines undermine the economy and food security in affected countries, and the expense of medical care and rehabilitation adds economic disability to the physical burden.

Concerning the economics of landmine clearance, Harris (2000) presented a cost-benefit analysis of landmine clearance in Cambodia and the alternative use of official development assistance resources. Harris arrived at the conclusion that costs are far greater than benefits in the low-income areas; therefore, efforts should be made in the areas of demining technology efficiency rather than demining itself. Finally, Paterson (2001) criticized the Harris model, conclusion, and policy recommendations. The author suggested a reassessment to assure better accountability and a more balanced approach to the clearance of agricultural and pasture land with high priority for the rural population. Mine clearance in extremely poor countries is justified in cost-benefit terms if the programs are well managed; for example, if productivity increases incrementally, if clearance efforts are well targeted, and if the programs are coordinated with rural development programs.

3. The Pseudo Panel Data Methodology

The landmine data used in this study is originally repeated cross-sectional data collected on a yearly basis. The observed individuals are those subjected to landmine accidents, and the data contains individual and environmental characteristics. Since the data primarily contains information about the accidents and little about post-accident measures, it provides little information useful for planning and allocation of resources. This is the reason for the attempt to transform the data into synthetic panel data, labelled pseudo panel data by the literature. This form of panel data allows for the capturing of both locational heterogeneity and temporal changes in the prevalence of landmine accidents.[3]

Panel data is constructed through pooling the observations from time-series data across different cross-sectional units, such as countries, regions, firms or households. It allows for capturing heterogeneity among the units observed and temporal patterns of their behaviour over time. In particular, panel data sets, in addition to the estimation of effects of observable variables, have the advantage that they can be used to identify and estimate unobservable unit-specific and time-specific effects. It gives a large number of degrees of freedom and more stable parameter estimates. However, the two dimensions of the data lead to more complex estimation procedures. Baltagi (2008) provides a comprehensive review and discussion of econometric estimation of panel data.

The data in our study is collected by the United Nations Office for Project Services Mine Action Program's (UNOPS-MAP) general survey teams covering mine-affected areas in the Kurdistan Region. The data set consists of 12,863 killed/injured observations. It covers the time period of 1970 to 2003 for twenty-six main districts in the Kurdistan region. A pseudo panel data is created to be able to specify a multivariate regression model to analyze the outcome of mine accidents. Here, the victims are grouped according to their time-invariant characteristics over time. Pseudo panel data is useful in cases where policy target groups of individuals are distinguished by their common time invariant characteristics, rather than as individuals. Moreover, grouping into cells will homogenize the individual effects so that the average district specific effect is approximately invariant between two periods, and if necessary for estimation purpose, is efficiently removed by within or by first difference transformations of the data (see Deaton 1985).

The pseudo panel data is created using the repeated cross sections where the outcome is panel data consisting of regions observed over time. Each observation represents average or cohorts of multiple observations for a region in a given year. It is common that one multiplies the average observation with the square root of the cell size to account for different intensities in the size effect. The resulting error component panel data model is:

(1) $$y_{ct} = \alpha_0 + \sum_{j=1}^{J} \beta_j x_{jct} + \varepsilon_{ct}, \quad where \quad \varepsilon_{ct} = \mu_c + \lambda_t + \nu_{ct}$$

where y is the outcome of accident for cohort *c* observed in time period *t*, x_{ct} is a matrix of *J* explanatory variables, β_j is a vector of *J* unknown parameters to be estimated, μ_c are unobserved individual effects, λ_t are unobserved time effects, and ν_{ct} are random error terms, and, following the tradition, it is assumed to have a mean zero and a constant variance. This specification allows for the standard within or least square dummy variable estimation of unknown parameters that are consistent and unbiased under the assumption of fixed effects. Since our data uses districts as the unit of observation and contains all districts, a fixed effects approach is suitable.

4. The Pseudo Panel Data

The data was collected mainly by UNOPS-MAP through conducting a general impact survey by visiting all mine-affected villages in the Kurdistan region. The MAP team interviewed the key persons in these villages and the victim's nearest relative or head of the family using a questionnaire form. The data set covers the period 1970 to 2003 for twenty-six main districts in the Kurdistan Region. It consists of 12,863 killed or injured individual observations. The data has not been updated by the regional mine action authority since 2003 when the UNOPS-MAP's activities were terminated during the coalition force invasion of Iraq. The data contains information about the outcome of the accidents in the form of the killed or injured and a number of individual and regional socio-economic characteristics such as age, gender, occupation, activity at the time of accident, knowledge about the danger, training program, location, and time of accident. In Table 10.1 below the detailed descriptions and definitions of the variables used in our analysis are presented. The summary statistics of the data used are presented in Table 10.2 in two parts: one is based on the original repeated cross-sectional data, and the other is based on the constructed pseudo panel data.

Table 10.1. Description of Variables used

A. Dependent variable:
Victim status after accident: Killed=1, Injured=0
B. Independent variables:
Age of the Victim: Years
Gender: Male=1, Female=0
Occupation of the victim: Peshmarga (pesh), child (child), unemployed (unempl), shepherded (sheph), farmer (farmer), driver (driver), others (othocc)
Activity at time of accident: Walking in the minefield (walk), touching the mine (touch), collecting the mines (collect), fighting in the minefield (fight), making a fire for cooking (fire), others (othact)
Knowledge about the danger: The victim knew that the area is dangerous (danger=1, no=0)
Training programs: The victim received a mine awareness training(mre: yes=1, no=0)
The incidents location: 26 districts, Akre, Amedi, Chamchamal, Choman, Darbandikhan, Dashtihawler, Duhok, Dukan, Erbil, Halabja, Kalar, Kiffri, Khanaqin, Koysinjaq, Mergasur, Penjwin, Pshdar, Ranya, Semel, Shaqlawa, Sharazur, Sharbazher, Shekhan, Soran, Suleimaniya, and Zakho, use 26 districts, district dummies, di1-di26.
Time of incident: 34 years, 1970–2003, use time trend (t) or 34 time dummies, dt1-dt34.

Table 10.2: Summary Statistics for the variables used

	Repeated cross sectional survey data, 12,863 observations					Pseudo panel data-not accounted for cell size, 738 observations				
Variable	Mean	Std Dev	Min.	Max.	t-value	Mean	Std Dev	Min.	Max.	t-value
male	0.912	0.283	0	1	365.8	0.927	0.135	0	1	187.35
female	0.088	0.283	0	1	35.19	0.073	0.135	0	1	14.66
age	27.578	16.406	1	100	190.74	27.733	10.164	3.50	77.00	74.13
killed	0.822	0.383	0	1	243.81	0.798	0.278	0	1	78.11
injured	0.178	0.383	0	1	52.80	0.202	0.278	0	1	19.72
pesh	0.133	0.340	0	1	44.46	0.178	0.242	0	1	19.94
child	0.240	0.427	0	1	63.71	0.215	0.250	0	1	23.35
unempl	0.076	0.266	0	1	32.64	0.065	0.128	0	1	13.73
sheph	0.129	0.335	0	1	43.65	0.157	0.233	0	1	18.39
farmer	0.383	0.486	0	1	89.47	0.338	0.278	0	1	32.99
driver	0.037	0.189	0	1	22.26	0.044	0.121	0	1	9.96
othocc	0.001	0.037	0	1	4.25	0.003	0.035	0	0.67	2.56
walk	0.288	0.453	0	1	72.15	0.270	0.264	0	1	27.73

touch	0.323	0.468	0	1	78.38	0.318	0.291	0	1	29.68
collect	0.263	0.440	0	1	67.74	0.265	0.280	0	1	25.70
fight	0.023	0.151	0	1	17.50	0.037	0.121	0	1	8.37
fire	0.018	0.134	0	1	15.54	0.016	0.070	0	1	6.14
othact	0.085	0.278	0	1	34.51	0.094	0.159	0	1	16.12
danger	0.240	0.427	0	1	63.82	0.228	0.273	0	1	22.72
mre	0.025	0.155	0	1	18.09	0.035	0.132	0	1	7.30
trnd	-	-	-	-	-	18.725	9.270	1.00	34.00	54.87

Note: t-value indicate probability of variable being different than zero.

In the cross-sectional data case, the victim occurrence time interval is 1929–2008, whereas in the pseudo panel the data is grouped based on time interval 1970–2003 because there were only a few victim records before and after these periods (see Fig. 10.1 below). The original data contains 12,863 observations, while the pseudo panel data is an unbalanced panel data with 738 observations. The difference is due to the fact that, instead of individuals in the cross-sectional case in the pseudo panel, we use the district of location as the unit of observation. The issue of maintenance of the property of the data retained and truncation of the time period can be checked and shown through comparing the average mean of the key variable across the two data sets.

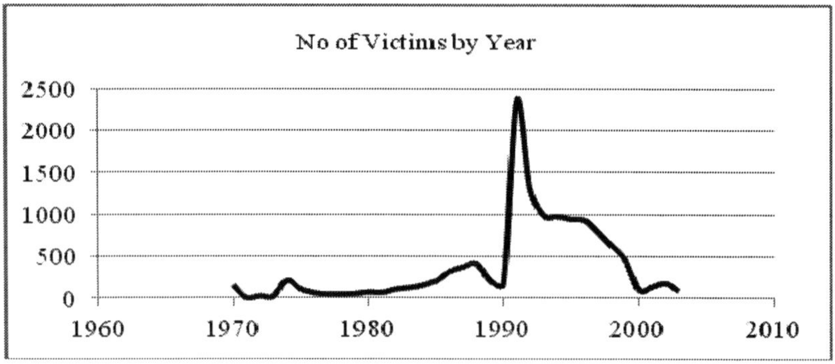

Fig. 10.1. Prevalence of landmine accidents over time in the Kurdistan region of Iraq.

The average age of the victim is 27.5 (27.7) years with 16.4 (10.1) years in dispersion around the mean age. About 82% (79.0%) of the victims were killed in the cross-sectional data case with dispersion of 3.8% (2.7%). The numbers in parenthesis are those of the pseudo panel data. The equality of age and killed rate following an accident indicates that the pseudo panel data construction and use of cell average do not lead to loss of cohorts' information. It shows that the properties of the parameter estimates are not changing noticeably because of the use of cohort means. We observe that the difference between the mean values of the variables in both scenarios is relatively small whereas the rate of dispersion is smaller in the pseudo panel. The pseudo model's fit measured by coefficients of determination or root mean squared errors is much better.

According to the pseudo panel data results, 79% of the victims are killed when being subjected to landmine accidents, in which only 2% were children. In total, 17% were working as Peshmarga soldiers, 33% as farmers, 15% as shepherds, and the remaining were unemployed, drivers, or other professionals. For the activity when the accident happened, 31% of the victims were touching the mine for the purpose of relocating it, 27% were walking inside the minefield, 26% were collecting the mine to be destroyed, and the remaining were either fighting, making fires, or engaged in other daily activities. The results show that only 22% of the victims knew that the area was dangerous, and only 3% of the victims had received mine awareness information and training programs.

The victim's status (killed=1 or injured=0) is used as the dependent variable. In the cross-sectional case, one must use qualitative logit or probit models to estimate the effects of determinants on the probability of getting killed or injured in the event of an accident (see Greene 2008). The problem here is that we do not have information about non-victims as the statistics cover only landmine victims, which causes a sample selection bias. In the pseudo panel case, the dependent variable is the share of victims killed, which is a continuous variable in the interval 0 and 1 where the level is ignored. Alternatively, one can use the aggregate number of killed and injured as a dependent variable. The aggregate number has the advantage that it accounts for levels and allows for the use of the log-linear functional form, easing direct interpretation of the estimated coefficients. The estimation result should not differ by using the share of killed or the count numbers of killed in accidents, given the same explanatory variables are used in the two model specifications. These two variable definitions allow for estimation of the models by using traditional regression models such as pooled, fixed or random effects. The model will estimate the effects of determinants on the probability of getting killed.

For explanatory variables, we used individual demographic characteristics and were able to distinguish the age and gender of the individual victims. Occupation is specified as: Peshmarga soldiers, child, unemployed, shepherd, farmer, driver and other occupation. Activity at the time of accident is divided into: walking in the minefield, touching the mine to destroy it, collecting the mines, engaging in conflicts taking place in the minefield, making fire for cooking, and other activities. We also have information about whether the victims knew about the dangers in the area or not, and whether the victims received a mine awareness training program or not. In addition to the above determinants, we have controlled for unobserved districts and time of accident effects by using twenty-six districts and thirty-four yearly time dummies. As an alternative to the use of yearly time dummies, we used a time trend and its square.

The majority of the victims (91.2%) in the data set are male. The average age is 27.6 years. Of the victims, 82.2% were killed and the remaining 17.5% were injured. Farmers (38.3%), children (24.0%), Peshmarga soldiers (13.3%) and shepherds (12.9%) are the main occupation groups of victims. Concerning the nature of the victims' activities at the time of accident, touching mines (32.3%), walking in the minefield (28.8%), and collecting mines (28.3%) are the main activities. Only 24.0% of the victims were aware of the danger, and only 2.5% had participated in information and awareness programs. These numbers are based on the original data of 12,863 landmine victim cases, which are close to their pseudo panel data counterparts mean values.

In a number of tables and graphs in Appendix C, we report the aggregate time series of the key variables by different characteristics of the individual victims. Development of the number of victims divided into killed and injured, the two categories' share of the total number of victims, and the share of killed distinguished by gender are reported in Appendix C Table C.A and further illustrated in Figs. C.A1, C.A2 and C.A3. The irregular numbers around 1977 to 1978 and 2000 to 2003 are very likely attributable to statistical errors and not necessarily to health care effectiveness or changes in landmine devastation effects. The variations in percentages of the killed across occupation groups and their development over time is reported in Appendix C Table C.B and their number and shares of killed illustrated in Figs. C.B1 and C.B2. The variations in percentages of killed across different activity groups and their development over time are reported in Appendix C Table C.C, and their number and shares of killed illustrated in Figs. C.C1 and C.C2. In a similar way, the variations in the number and percentages of killed who were aware of the danger and participated in mine awareness programs and their development over time are illustrated in Appendix C Figs. C.D1 and C.D2.

Table C.D shows the number of killed and injured across different districts, demonstrating large variations in the size of the population. In general, there is a negative relationship between the population density and frequency of accidents. The number of killed per 1,000 population in the two major cities of Erbil and Suleimaniya is, relative to other districts, low, being 0.27 and 1.12, respectively. The numbers are far below those of Derbandikhan (10.17), Pemjwin (20.72), Choman (20.94), and Sarbazher (50.83). The ratio of killed to injured is large, indicating low access and the long distance to professional health care centres within these districts.

5. The Model and Estimation Procedures

We have constructed four models to explain variations in the share of killed victims by information obtained on socio-economic characteristics of the individuals. These models are fixed effects and formulated as follows.

Model 1 is a fixed effects time dummy model where districts and year dummies are included to capture unobserved time-specific and district-specific effects. The variables are average observations in a district for a given year. No adjustment is made for the cell (district) size or frequency of victims for a district in a given year. The frequency of victims varies by district and over time. The overall frequency of accidents or development of the number of victims over the period 1970 to 2003 is shown in Fig. 10.1 above.

Model 2 is a fixed effects time dummy model that accounts for district-specific and time-specific effects. The difference with Model 1 is that the variables are adjusted for the frequency of accidents in a district in a given year. The model accounts for size of cell by allowing for its impact on the determinants of the accident rate and the outcome but at a decreasing rate, because the square root of the number in the cell is used in the multiplication with the variables.

Model 3 and 4 correspond to Model 1 and 2, but the difference is that in these two models, instead of using yearly time dummies to capture unobserved time-specific effects, we include a time trend and its square. This reduces the number of time-specific effects from 34-1 to only 2, excluding the intercept representing the reference year (1970). Despite its non-linearity in time, it loses power, which comes at the cost of losing year-to-year variations, but on average the two models should provide somewhat similar trends. In our view, the time dummy model should be preferred because the frequency distribution over time (see Fig. 10.1) suggests the time trend is ineffective in capturing the year-to-year variations.

Multicollinearity is the condition where the independent variables are linearly related to each other. A model with high correlation between two or more independent variables is said to have a multicollinearity problem (Wooldridge 2006). Through specification of our models, we incorporate explanatory variables that are highly correlated with the dependent variable, but we choose explanatory variables that are not correlated with each other to avoid mixing the effects estimated in the form of coefficients. Correlation coefficients between 0 and 0.3 are considered weak, those between 0.3 and 0.7 are moderate, and those between 0.7 and 1.0 are considered high. Therefore, it is desirable that the interpretation of the effects of explanatory variables in the regression results is reliable and not confounded. In general, in textbooks, ±0.50 is used as the threshold in the evaluation of multicollinearity. However, an acceptable interval of correlations reported by Wheeler and Tiefelsdorf (2005) is somewhat higher, ±0.59.

In all the models under study, correlation coefficients between the independent variables that are reported in Table 10.3 are less than ±0.54 and statistically significant mostly at less than a 1% level of significance, which indicates that multicolinearity is not a serious problem in this case. Some of the explanatory variables are positively correlated, while others are negatively correlated with all explanatory variables, including age, gender, occupation, activity, knowledge and trend. The degree varies in the interval -0.53 until +0.10. The highest negative value is age, while the highest positive occupation is farmer in relation to the dependent variable. Some of the independent variables are also positively correlated, while others are negatively correlated. As mentioned previously, the degree of correlation here is lower than the threshold for the serious degree of collinearity. The highest correlation coefficient is -0.54, which is attributed to unemployed-male combination. A few other negative correlations are child-age (-0.36), walk-touch (-0.41), collect-touch (-0.041), and collect-walk (-0.37) related to the nature of activities.

Estimation results are reported in Table 10.4 below. As mentioned previously, four models are estimated in which they differ according to whether, in the construction of the pseudo panel, one accounts for the frequency of accidents or uses simple averages by districts, and whether the time-specific effects are captured by yearly time dummies or a time trend and its square.

The F-test is a statistical test used to evaluate the null hypothesis that all slope regression coefficients are equal to zero versus the alternative that the intercept is not zero. A significant F-test indicates that the observed R-squared is reliable and is not a spurious result of oddities in the data set. Thus, the F-test determines whether the proposed relationship is statistically reliable. The F-tests in the four estimated models are all statistically significant at less than a 1% level of significance, suggesting that the model specification is adequate. The Bayesian Information Criterion (BIC) and Variance Inflation Factors (VIF) also suggest that the models are not being over-fitted and are specified correctly. The standard errors are robust standard errors corrected for heteroscedasticity.

Table 10.3: Pearson Correlation Coefficients (738 Observations)

	killed	male	age	pesh	child	unempl	sheph	farmer	driver	othocc	walk	touch	collect	fight	fire	othact	danger	mre	tmd
killed	1	-0.013	-0.532	-0.068	-0.091	0.091	0.013	0.103	-0.006	-0.094	0.047	-0.036	-0.040	-0.029	-0.010	0.083	-0.151	-0.176	-0.110
male	-0.013	1	-0.024	0.133	-0.077	-0.537	0.169	0.028	0.064	0.023	-0.017	-0.010	0.085	0.012	-0.142	-0.051	0.023	0.070	0.056
age	-0.532	-0.024	1	0.154	-0.362	0.012	-0.090	0.202	0.127	0.037	0.150	-0.175	-0.004	0.098	0.017	-0.004	-0.058	0.004	-0.050
pesh	-0.068	0.133	0.154	1	-0.269	-0.133	-0.242	-0.306	-0.116	-0.067	0.251	-0.240	-0.178	0.298	-0.061	0.137	-0.193	-0.039	0.001
child	-0.091	-0.077	-0.362	-0.269	1	-0.083	-0.222	-0.388	-0.112	-0.037	-0.112	0.306	-0.155	-0.149	0.037	-0.005	0.088	0.032	0.057
unempl	0.091	-0.537	0.012	-0.133	-0.083	1	-0.134	-0.130	-0.058	-0.017	0.068	-0.030	-0.050	-0.051	0.141	0.008	-0.001	-0.073	0.027
sheph	0.013	0.169	-0.090	-0.242	-0.222	-0.134	1	-0.322	-0.094	-0.008	-0.227	0.032	0.306	-0.130	-0.041	-0.102	0.110	0.020	-0.014
farmer	0.103	0.028	0.202	-0.306	-0.388	-0.130	-0.322	1	-0.125	-0.009	0.053	-0.063	0.062	-0.064	0.002	-0.034	-0.051	-0.058	-0.073
driver	-0.006	0.064	0.127	-0.116	-0.112	-0.058	-0.094	-0.125	1	-0.025	-0.006	-0.028	-0.026	0.169	-0.029	-0.009	0.065	0.141	0.023
othocc	-0.094	0.023	0.037	-0.067	-0.037	-0.017	-0.008	-0.009	-0.025	1	-0.070	-0.027	0.090	-0.029	-0.003	0.032	0.156	0.150	0.074
walk	0.047	-0.017	0.150	0.251	-0.112	0.068	-0.227	0.053	-0.006	-0.070	1	-0.414	-0.372	-0.101	-0.089	-0.133	-0.332	-0.143	-0.178
touch	-0.036	-0.010	-0.175	-0.240	0.306	-0.030	0.032	-0.063	-0.028	-0.027	-0.414	1	-0.440	-0.187	-0.066	-0.199	0.335	0.029	-0.032
collect	-0.040	0.085	-0.004	-0.178	-0.155	-0.050	0.306	0.062	-0.026	0.090	-0.372	-0.440	1	-0.117	-0.054	-0.221	0.014	0.129	0.163
fight	-0.029	0.012	0.098	0.298	-0.149	-0.051	-0.130	-0.064	0.169	-0.029	-0.101	-0.187	-0.117	1	-0.056	-0.017	-0.104	-0.019	0.021
fire	-0.010	-0.142	0.017	-0.061	0.037	0.141	-0.041	0.002	-0.029	-0.003	-0.089	-0.066	-0.054	-0.056	1	-0.030	0.026	0.065	0.077
othact	0.083	-0.051	-0.004	0.137	-0.005	0.008	-0.102	-0.034	-0.009	0.032	-0.133	-0.199	-0.221	-0.017	-0.030	1	-0.020	-0.057	0.018
danger	-0.151	0.023	-0.058	-0.193	0.088	-0.001	0.110	-0.051	0.065	0.156	-0.332	0.335	0.014	-0.104	0.026	-0.020	1	0.268	0.245
mre	-0.176	0.070	0.004	-0.039	0.032	-0.073	0.020	-0.058	0.141	0.150	-0.143	0.029	0.129	-0.019	0.065	-0.057	0.268	1	0.323
tmd	-0.110	0.056	-0.050	0.001	0.057	0.027	-0.014	-0.073	0.023	0.074	-0.178	-0.032	0.163	0.021	0.077	0.018	0.245	0.323	1

Table 10.4: Fixed effects estimation results, n=738 obs

	Model 1: fixed effects, time dummy, no cell size adjustment		Model 2: fixed effects time dummy, with adjustment for cell size		Model 3: fixed effects, time trend, no cell size adjustment		Model 4: fixed effects time trend, with adjustment for cell size	
F-test	21.42***		181.35***		21.10***		274.47***	
Adj R^2	0.6721		0.9477		0.5398		0.9410	
RMSE	0.1590		0.5184		0.1884		0.5503	
Variable	Coefficient	Std Err1	Coefficient	Std Err1	Coefficient	Std Err1	Coefficient	Std Err1
Intercept	1.0861***	0.0951	0.2930**	0.1239	1.1724***	0.1074	0.0379	0.1158
male	0.0480	0.0604	0.9188***	0.0561	-0.0114	0.0691	0.9725***	0.0611
age	-0.0047	0.0033	-0.0134***	0.0016	-0.0092***	0.0039	-0.0137***	0.0016
age^2	-1.47E-4***	5.08E-5	1.82E-6	3.37E-6	-1.04E-4*	6.13E-5	5.19E-8	3.08E-6
child	-0.1422***	0.0513	0.1312**	0.0625	-0.1937***	0.0525	0.0955	0.0650
unempl	0.1314**	0.0694	1.0865***	0.0894	0.1284*	0.0767	1.0987***	0.0910
sheph	-0.0097	0.0540	0.2824***	0.0755	0.0127	0.0644	0.3359***	0.0822

farmer	0.1324***	0.0425	0.5137***	0.0574	0.1113***	0.0462	0.5155***	0.0604
driver	0.1522***	0.0578	0.4370***	0.1028	0.1425**	0.0749	0.3803***	0.1226
othocc	0.3590**	0.1709	0.4566**	0.2115	-0.1989	0.2263	-0.2222	0.2268
touch	-0.0475	0.0411	-0.1011**	0.0490	0.0276	0.0478	-0.0433	0.0538
collect	0.0651	0.0409	-0.0524	0.0567	-0.0548	0.0472	-0.1036*	0.0623
fight	-0.0993	0.0720	0.1091	0.1180	-0.0945	0.0748	0.1349	0.1327
fire	0.0469	01081	0.0230	0.1157	0.1122	0.1443	0.1535	0.1357
othact	0.0338	0.0497	0.0978	0.0680	0.0872*	0.0514	0.1472**	0.0704
danger	0.0234	0.0389	0.0127	0.0435	-0.1319***	0.0506	-0.0471	0.0476
mre	-0.0273	0.0926	-0.0426	0.1008	-0.1835	0.1328	-0.1006	0.1173
di2 (distr.)	-0.0043	0.0453	0.0620	0.1327	0.0141	0.0563	0.0155**	0.1344
di3	0.0660	0.0444	0.1419	0.1349	0.0448	0.0556	0.0656	0.1572
di4	-0.0495	0.0493	0.0379	0.1547	-0.0236	0.0563	-0.0118	0.1563
di5	0.0355	0.0437	0.0307	0.1195	0.0585	0.0548	0.0036	0.1187
di6	-0.2429***	0.0575	-0.5298***	0.1553	-0.2561***	0.0677	-0.5967***	0.1616

di7	-0.0753	0.0536	-0.1158	0.1368	-0.0447	0.0620	-0.1191	0.1480
di8	-0.1931***	0.0537	-0.6327***	0.1697	-0.1463**	0.0627	-0.6463***	0.1569
di9	-0.0955**	0.0537	-0.1248	0.1477	-0.1158**	0.0610	-0.1962	0.1611
di10	0.0583	0.0506	0.1875	0.1450	0.0858	0.0562	0.1456	0.1447
di11	0.0507	0.0485	0.1119	0.1357	0.0697	0.0624	0.0772	0.1333
di12	0.0769*	0.0449	0.1148	0.1594	0.0350	0.0690	0.0706	0.1887
di13	0.1021*	0.0608	0.0760	0.1300	0.1991***	0.0800	0.1277	0.1419
di14	0.0451	0.0476	0.1730	0.1204	0.0828	0.0579	0.1417	0.1251
di15	0.0236	0.0479	0.1227	0.1648	0.0238	0.0654	0.0964	0.1692
di16	0.0293	0.0505	0.1491	0.1278	0.0635	0.0608	0.1506	0.1420
di17	-0.1615***	0.0503	-0.6230***	0.1585	-0.1186**	0.0614	-0.6253***	0.1572
di18	-0.1881***	0.0543	-0.6401***	0.1816	-0.1512***	0.0622	-0.6472***	0.1807
di19	-0.0503	0.0735	-0.1985	0.1451	-0.0487	0.0772	-0.1513	0.1451
di20	-0.1371***	0.0517	-0.2302*	0.1374	-0.1125*	0.0631	-0.2659**	0.1409

di21	0.0300	0.0485	-0.0208	0.1234	0.0321	0.0608	-0.0302	0.1317
di22	0.0012	0.0458	0.1331	0.1369	0.0139	0.0565	0.0925	0.1441
di23	0.1177**	0.0603	0.1910	0.1292	0.1032	0.0758	0.1740	0.1421
di24	-0.0782*	0.0459	-0.2174	0.1562	-0.0633	0.0536	-0.3632**	0.1532
di25	-0.0355	0.0466	0.0472	0.1454	-0.0021	0.0587	-0.0277	0.1523
di26	0.0101	0.0503	-0.0514	0.1375	-0.0123	0.0522	-0.1291	0.1311
dt2 (year)	0.1035	0.0721	0.0116	0.1461	-	-	-	-
dt3	-0.0496	0.0612	-0.1071	0.1448	-	-	-	-
dt4	-0.1660**	0.0802	-0.4122***	0.1593	-	-	-	-
dt5	-0.0433	0.0312	-0.0628	0.1297	-	-	-	-
dt6	-0.0325	0.0347	-0.1981*	0.1126	-	-	-	-
dt7	-0.0701	0.0469	-0.2330*	0.1315	-	-	-	-
dt8	0.0012	0.0496	-0.1924*	0.1184	-	-	-	-
dt9	-0.1261**	0.0534	-0.3086**	0.1383	-	-	-	-
dt10	-0.0140	0.0493	-0.0454	0.1619	-	-	-	-
dt11	-0.0817*	0.0470	-0.1165	0.1104	-	-	-	-

dt12	-0.0442	0.0426	-0.0694	0.1200	-	-	-	-
dt13	-0.0051	0.0306	-0.0102	0.0987	-	-	-	-
dt14	-0.0146	0.0316	-0.1039	0.1020	-	-	-	-
dt15	-0.0242	0.0340	-0.0592	0.1301	-	-	-	-
dt16	-0.0936***	0.0334	-0.2286**	0.1118	-	-	-	-
dt17	-0.0399	0.0297	-0.1059	0.1074	-	-	-	-
dt18	-0.0337	0.0297	0.0014	0.1169	-	-	-	-
dt19	-0.1041**	0.0437	-0.2005*	0.1247	-	-	-	-
dt20	-0.0505**	0.0267	-0.1208	0.0984	-	-	-	-
dt21	-0.0748***	0.0308	-0.2574**	0.1080	-	-	-	-
dt22	-0.0727***	0.0282	-0.2554	0.2257	-	-	-	-
dt23	-0.0640***	0.0270	-0.1817	0.1410	-	-	-	-
dt24	-0.0735***	0.0278	-0.2228*	0.1384	-	-	-	-
dt25	-0.0688***	0.0283	-0.0756	0.1407	-	-	-	-
dt26	-0.0647**	0.0322	-0.0166	0.1536	-	-	-	-
dt27	-0.0590**	0.0300	-0.0688	0.1662	-	-	-	-
dt28	-0.0509**	0.0264	-0.0775	0.1237	-	-	-	-

dt29	-0.0763**	0.0370	-0.0773	0.1564	-	-	-	-
dt30	-0.0746**	0.0364	-0.1685	0.1219	-	-	-	-
dt31	-0.0812	0.0556	-0.1952	0.1321	-	-	-	-
dt32	0.0597	0.0434	0.0212	0.1366	-	-	-	-
dt33	-0.1836**	0.0798	-0.6398***	0.1981	-	-	-	-
dt34	-0.8430***	0.0660	-1.5300***	0.1913	-	-	-	-
Trend	-	-	-	-	0.0079**	0.0035	0.0189**	0.0094
$Trend^2$	-	-	-	-	-0.0003***	0.0001	-0.0008***	0.0003

Notes: Peshmarga, walking, Akre and 1970 are reference occupation, activity, district and years. The asterisks ***, ** and * indicate significant at the less than 1%, 1-5% and 6-10% levels of significance, respectively. 1. Heteroscedasticity consistent-standard errors.

By looking at the adjusted R-square for Model 1 (0.672) and Model 3 (0.539), where the variables are not adjusted for the cell size, this shows that only 67.2% and 53.9%, respectively, of the variations in the dependent variable killed can be explained. By looking at the estimated coefficients in Model 4, where the fixed effects of location and time trend are considered and the variables are adjusted for the cell size, we notice that all the variables are statistically significant at a 1% level of significance. The adjusted R-square is 0.941, implying that 94.1% of the killed data is well explained by this model. The R-square in the model, using yearly time dummies and accounting for frequency of the killed at the district level (Model 2), is 0.947. Thus, the explanatory power of Model 2 is highest among the four estimated models.

In this case, we cannot select our preferred model simply based on the coefficient of determination. A Chow test was conducted to compare the time dummy vs. time trend models, Model 2 vs. Model 4. The test result (F=1.373) suggests that the preferred model is Model 4, which accounts for frequency of accidents in a given district and use of a time trend to capture the unobservable time-specific effect. In comparing the time trend and time dummy models, we notice that the remaining parameters' significance level, sign and size seem to be robust and not affected by the specification of time-specific effects. Furthermore, the two models are not exactly nested and, as such, not testable using the Chow test. Due to these two facts, we should base our analysis on Model 4. However, we prefer the time dummy because the two models are not exactly nested, and therefore Model 2 is considered the preferred model specification.

6. Interpretation of the Results

The basic model estimated by ignoring district and time effects (not reported here) produces R-square of 0.467. Chow tests of the four reported models vs. the basic model show that the former models are preferred. The sole introduction of year dummy variables in Model 1 without adjusting for the cell size compared with Model 2 (adjusted) shows that the percentage of explained variations in the dependent variable rise from 67.2% to 94.7%. This has another implication, namely confirming that it is important to account for the frequency of the accident rate among the districts.

The result shows that the male effect is statistically significant in both Model 2 and 4, suggesting that males are more often subject to landmine accidents than their female counterparts due to the nature of their work. When adding the square of age to account for non-linear relationship between accident rate and age, we find age to be statistically significant in both Model 2 and 4 at less than a 1% level of significance. The squared term coefficient is insignificant, suggesting a linear relationship between accident rate and age of the victims. A higher age is associated with higher probability of being killed. However, the effect of age varies when district effects are controlled.

Estimation results based on Model 2 suggest that children, compared with Peshmarga soldiers as a reference group, are more prone to landmine accidents. The opposite is observed in Model 3. Similarly, the unemployed, shepherds, farmers, drivers, and other professions face landmine accidents more often than Peshmarga soldiers. The coefficient of the activity of touching mines is negative and significant, suggesting that walking is the main activity leading to fatalities. The remaining activities are not statistically different from the reference variable of walking. Unlike in the case of Husum et al. (2002), the data does not allow for post-accident treatment, rehabilitation, and evaluation of chronic pain in landmine accident survivors.

Only in Model 3 do we notice that information about the danger of landmine reduces the rate of accidents. Contrary to our expectations, participation in landmine training and awareness programs does not reduce the rate of accidents. This might be explained by the low frequency and/or ineffectiveness of such programs. Low rate of literacy in the rural areas might be another factor explaining the low training and awareness program effect.

District-specific effects, which are constant over time, are captured in the form of fixed effects. Akre is used as the reference district. Compared with Akre, the coefficient of five districts in Model 2 and six districts in Model 4 are statistically different from Akre. The negative sign suggests a lower accident rate than Akre. Dashti Hawler, Dukan, Peshdar, and Ranya have a lower accident rate perhaps due to low landmine concentration in these areas. Only nine of the yearly time dummy variables are found to be statistically different than that of the reference year 1970. The rate is lower in all nine years compared with 1970. In regards to the alternative specification with time trend and its square, we notice that in Model 3 and 4 both coefficients are negative and statistically significant. The signs suggest that the accident rate is declining at an increasing rate, which is contrary to the insignificant training program effects.

7. Socio-economic, Psychological, and Health Impacts

The socio-economic impact of landmines and UXO extends across multiple areas, such as access to public areas, salvaging metal from mines and UXO, accessing land for agriculture, and repairing existing infrastructures. The impact of landmines on socio-economic aspects can be either direct or indirect. The indirect impacts of landmines include malnutrition/starvation and the spread of infectious diseases because of the inability to repair water purification systems and to provide public health services in areas isolated by mines and UXO.

Our data does not fully allow for assessment of social, psychological and economic impacts of landmine accidents. The literature in this area is not developed, but there are examples of studies covering these important aspects. For instance, Anderson et al. (1995) investigated the effects of landmines on the health and social conditions of communities in four countries (Afghanistan, Bosnia, Cambodia and Mozambique). The focus was on the effects of landmines on food security, residence, livestock, land use and extent of injuries, as well as the physical, psychological, social, and economic costs of injuries. Landmines were found to undermine the food security and economy of the affected countries. Kakar et al. (1996) also found landmines to constitute a major public health problem in the world, suggesting its production ban.

In order to know the extent of the impact of landmine and UXO on social, health and economic sectors, it is worth mentioning that planting one mine costs US $3–15, while the process of removing it, ignoring the risk, costs US $100–350. According to UN estimates, the cost of treating an injured person as a result of the explosion of a mine is more than three times the cost of treating a person injured by other means.[4] Medical reports indicate that a mine victim, if not killed, needs at least three surgical operations in addition to further medical supplies. The minimum cost of treating a mine victim is US $6,000; each prosthetic limb costs US $3,000–5,000 and needs to be replaced for the victim every six months. Hence, the effects of landmines go far beyond killing, injuring, and disabling humans to affecting the nation's economy and individuals' well-being (Coupland 1996). In recent years, the sharing of oil revenues between regional and central governments in Iraq in proportion to the share of population allowed in the building up of health care capacity (Al-Bazzaz 2010) and the conducting of landmine treatment and rehabilitation (Shabila et al. 2010). However, due to the public and not-for-profit nature of these services, they are underinvested in and underdeveloped.

The economic sector that is mostly harmed by the presence of landmines in the Kurdistan region is the agriculture sector. The presence and threat of landmines often disrupt agricultural development and consequently the health and well-being of the rural population. There are several effects from landmines on agriculture throughout, impacting the amount of arable land for cultivation. Most of the fertile lands are mined, which in turn affects the supply of food and limits employment opportunities. The mined grazing lands limit animal husbandry activities, which are traditionally of significant importance for food supply. These landmines have not only stalled the agriculture sector but have also killed large numbers of livestock, the main source of livelihood in the rural areas. Landmines not only obstruct the development of agriculture, they also hinder economic development in general as a result of reduced investment. In line with Harris (2000) and critical commentary by Paterson (2001), one should conduct a cost-benefit analysis of landmine clearance of agricultural land to justify resource allocation for clearance activities, treatment and rehabilitation. However, many of the positive social effects of such investments are not possible to measure quantitatively and monetarily.

In addition to the physical impact and the socio-economic consequences of landmine injuries, they have severe emotional and psychological impacts on the victims' families. The emotional status of landmine victims is important because disabled individuals often turn to alcohol and acquire aggressive behaviours, which result in serious public health and social problems. Thus, the mere threat of the existence of landmines, let alone their actual presence, disrupts everyday activities, such as fetching safe water and cultivating fields for food production. It also deters public health campaigns and dramatically alters the social and economic state of the injured and their families, ultimately affecting the structure and well-being of entire societies. Accounting for all of these negative effects and externalities, they were expected by the central government's armed forces who placed the mines in the first place, with the aim to depopulate the region and to effectively destroy the livelihood of the citizens who were demanding their constitutional and international human rights.

Landmines continue to maim and kill, also causing mental health consequences for victims and their families. In addition, small arms and landmines cause a substantial burden on the health sector as well as the economic condition of the families of the victims. The economic costs associated with injuries from small arms and landmines are estimated to be in the billions of dollars each year.[5] These consume a large share of public and private budgets, reducing the states' and households' potential for development and prosperity.

8. Summary, Conclusion and Policy Recommendations

Landmines and UXOs are aimed for use in war or conflicts between countries for defence and security reasons. However, their use has frequently been abused by some national governments against rebelling minorities with the purpose of depopulating certain regions in order to change the demography in an attempt to monopolize natural resources. Thus, natural resources are a double curse for minorities.

This chapter analyzed the data of victims caused by mines and UXOs in the Kurdistan Region of Iraq. The data is yearly cross sections collected from the late 1960s until 2003. In addition to descriptive analysis, we use regression analysis to identify determinants of the probability of death by a landmine and to estimate the individual determinants' impacts. The individuals are observed when they are subjected to landmine accidents. In order to account for the dynamic aspect of the process and heterogeneity by location, we created pseudo panel data where each of the districts is observed over time for the entire period. This allows for analysis of both heterogeneity and dynamics across locations and over time, as well as to control for unobserved location and time effects. In our view, the result based on the pseudo panel data is more useful in monitoring, planning and resource allocation.

Empirical results show that males are more often subjected to landmine accidents than their female counterparts, which is attributed to the nature of their activities. There is a negative age effect, suggesting younger people are less exposed to the risk of accidents. The probability of accident is lower for children, unemployed individuals, shepherds, farmers, and drivers compared to Peshmarga soldiers. Walking is the type of activity that leads to a higher rate of accidents compared with touching mines and other daily activities. Awareness of landmine danger reduces the probability of accident, but no significant reduction is observed from participation in training programs.

Concerning the unobserved time and district effects, we noted that in some districts and years the accident rate is higher than in 1970 and in Akre, which serves as a reference year and district, respectively. A major part of the decline in the rate of accidents is attributed to urbanization and migration from agriculture to service and industry sectors. Despite these major changes, the population is highly exposed to the disproportionate danger of landmines harming the economy and social life, and causing enormous psychological suffering and pain for the entire population.

The Kurdistan regional government is urged to allocate a fixed share of its annual budget raised from oil and gas revenue sharing to landmine awareness and clearance, health care capacity and quality, and rehabilitation programs. This will make the region independent of foreign NGOs, which might take longer than expected to clear the area due to other priorities around the world. Planning and management of mine action, health care and rehabilitation in the Kurdistan region should be strictly in the hands of local people. Currently, the stalled mine action group's activity is in the hands of the central government, sections of which in the past actively participated in the minelaying in the region. As such, a centrally nominated and financed authority is certainly not trustworthy to lead the clearing actions and related health and rehabilitation activities. Judging from the recent years of high public revenues, there is no shortage of funds to reactivate the landmine action group in the region in order to cope with this hazardous factor that harms the general population.

A research institute should be established to conduct research on landmines and collect statistics. Such an institute should play a major role in the coordination of research related to different aspects of landmines, such as information, prevention, training and rehabilitation, the labour market and pensions. In addition, it should enable the gathering of evidence on crimes committed and encourage and support the regional governments in following up on the landmine issue in matters of prevention, impact reduction, victim compensation and punishing criminals.

Notes

[1] International Campaign to Ban Landmines, the Landmine Monitor Report 2009. Available at http://www.icbl.org, Accessed on Dec. 21, 2010

[2] U.S. Department of State, *Hidden Killers: The Global Problem with Uncleared Landmines*, 1993; Hidden Killers: The Global Landmine Crisis, 1998.

[3] For discussion of the properties of pseudo panel data and their advantages and limitations, see Verbeek (1992), Verbeek and Nijman (1992); for their applications, see Heshmati and Kumbhakar (1997).

[4] http://www.un.org/cyberschoolbus/banmines/units/unit1c.asp, Landmine-Related Injuries (1993–1996), Morbidity and Mortality Weekly Report 1997, 46, 724–726, Accessed on Dec. 11, 2010.

[5] http://www.picosearch.com/cgi-bin/ts.pl?index=159730&calln=6&lastq=&sortsel =rel&psel=jma&opt= ANY&query=health%20impact, Accessed on March 1, 2011.

CHAPTER ELEVEN

HUMAN SECURITY AND THE POLITICAL IMPACTS OF LANDMINES IN THE KURDISTAN REGION

Landmines and UXOs can be considered as a serious threat to the development and reconstruction of societies. They have strong negative effects even on the process of integration within the society. This is obvious in politically unstable communities in particular that prevail the mechanisms of narrow expression in order to obtain rights and benefits.

People from the third world mostly pay the bill of mines twice compared with their counterparts in the developed nations for two reasons: first in trying to obtain mines at the expense of satisfying the needs of education, health, and development, and second in trying to get rid of them and avoid their negative effects. The later cost component is much higher than the first, involving danger to life attributed to the risky but necessary cleaning landmines activities.

The Kurdistan region of Iraq was not excluded from the phenomenon of mines and UXOs and their deep effects. This is due to the extension of continuous conflicts and wars throughout the political history of the region, which in turn increased the depth of the case of its negative effects. Over time, these negative effects extended to include most of the communities in the region. The intensity of landmine and UXO prevalence differs though from one community to another from the military perspective due to the strategic location of these communities.

The landmines and UXOs continue to threaten the lives of innocent people, with millions scattered throughout the region as a consequence of years of inter-country wars and internal conflicts. The civilians are the most frequent victims of mines and UXOs in war and even in time of peace. Subsistence farmers, refugees, villagers cultivating their fields, herding animals, collecting firewood and bringing water are forced into often dangerous, mined areas in order to access basic resources to satisfy their daily survival.

Injuries caused by mines and UXOs affect the lives of the victims as well as their families and the medical infrastructure of the country. The landmine victim most likely will not be able to earn income except for the famers who cultivate contaminated lands that have no other choice but to support their life and their families. The landmine victims' ability to have a standard living style will notably be shortened beside the psychological damage and consequences for those victims who suffer physical injuries such as limb amputation. The psychological impacts of landmine accidents differ for males and females. It is obvious that the impact on female victims is more severe than on male, and it effectively reduces their chances to find a job, get married, and to have children and a somewhat normal social life.

The landmine victim injuries heavily affect the contaminated country's medical infrastructure. The types of wounds are more likely expected to be infected and need a series of surgeries and treatments. For the case of amputees, they need wheelchairs or prosthetic limbs and intensive physiotherapy to learn to use them. The artificial limbs will need to be replaced at regular periods as they become deteriorated, and in the case of children the replacement is even more frequent as they grow, with limbs needing replacement several times each year. All of these impose significant pressure on the already limited health services available.

The impact of landmines and UXOs extends far beyond individual accidents and casualties. Huge areas of former battle fields have created acute social, economic and environmental harm through denying access to productive land areas for civilian use. Landmines and UXOs have a negative impact on development, and it obstructs post-war rebuilding activities, denying access to health and education centres. Many agriculture lands have become useless for this reason. Mines have been placed in irrigation systems and wells making it impossible for farmers to cultivate their lands, which is their only source of income, and shepherds to herd animals freely and where appropriate for the purpose.

Water dams and power stations with the presence of mines and UXOs reduce the ability of the government to produce the required power for industry, households and the well-being of the nation. The transportation system affected by mines and UXOs interrupts the movement of people and the transportation of goods and services. NGOs working with humanitarian assistance face difficulties in reaching affected communities and areas due to mined roads and destruction of infrastructures.

In sum, landmines and UXOs cause massive collateral damage to the society. They are different from all other weapons due to their indiscriminate nature and persistent threat after wars. Lack of accountability, responsibility and public service provision often lead to their continuous threat in the rural areas.

Hence, there is an urgent need to reduce the impact of landmines and UXOs on the people and their families and on the communities so that they can live free from the threat and impact of landmines. Such decisions are made by armed forces authorized by urban politicians. The shift of the burden to people who do not gain from cross border or internal conflicts is highly unfair.

The rest of this chapter is organized around explanations of the multi-dimensionality of human security. In the next sections, we discuss human security needs encompassing economic, food, health, environment, personal, community and political aspects.

1. Human Security

Discussions about human rights in the third world cover mostly the political, civil and economic aspects of human rights. However, other types of rights have been under debate in recent years. These include the right to develop and to have peace, which it is a serious challenge facing the developing countries. The aspect of right to walk freely without fear can be considered the most important issue for the right for peace. The presence of landmines and UXOs do not allow the people to return to their villages, to resume their agriculture activities, shepherd their livestock, access water and enjoy nature. In addition, the fear, anxiety and psychological impacts of landmines and the high risk conditions cause a great deal of problems for parents regarding the life of their children and other family members.

Human security is a buzzword that is frequently used in many contexts in recent decades, such as in relation to sustainability in population growth, economic growth, security, exhaustible resource utilization and management of outbreak of disease of global nature that threatens humankind and our environments. Alkire (2003) views human security as a model to understand the weaknesses of global supporters to challenge the existing traditional concept of national security, by saying that the correct reference for security should be the individual and not the state.

The concept of human security focuses on the viewpoints of individuals to protect them from threats to human lives, livelihoods and dignity, and to bring out the full potential of each individual. Human security which carries a view point of individual centred security is essential to achieve national stability at the regional and global levels. This concept has emerged as a means to understand the post-Cold War multidisciplinary security, and includes a number of research areas such as development studies, international relations, global health and strategic studies, and human rights (Bajpai, 2000).

The United Nations Development Program (UNDP) 1994 Human Development Report is a landmark publication in the field of human security, and it secures the argument that "freedom from want" and "freedom from fear" for all people is the best way to address the problem of global insecurity. It is referred to in a wide range of global policy discussions and scientific journals and is often taught in universities, human security being part of international relations, globalization and human rights studies.[1]

According to the report, human security should be expanded to include global security threats in seven areas: economic, food, health, environment, personal, community and political security aspects. Each of these aspects is explained below in a more detailed form.

1.1 Economic Security

Economic security requires a basic guaranteed income to be provided to individuals for the work to be productive and rewarding, or as a last resort as a safety net funded by the public sector. In this regards, only about a quarter of the world's population for the time being are economically secure. While the problem of economic security may be more serious in developing countries, concerns also arise in developed countries, with problems of unemployment often being an important factor behind political tension and ethnic violence.

The economic impact of landmines and UXOs can be seen as one of the determinant factors of economic security, as it prevents people in the mines and UXO affected areas from working. In addition, the surviving mine victims with disabilities are not be able to be employed in most cases and are unable to earn money without being depending on their families, charity organizations or the state.

Bier (2003) mentioned that the most important effect of mines and UXOs is on economic development. The area of negative impacts is large, and here we limit ourselves to a few most frequent examples. They destroy all means of the living and working environment, such as livestock and flora and fauna, dislodges the market and the market production, hampers the investment potential, especially for tourism, slows down the provision of government services, thwarts the efficient use of arable lands, drains the medical services and public health campaign, increases instability due to the lack or delay of economic development, hampers the transportation of goods and services and prevents export and import trades, hampers the internally displaced persons return to their homes, prevents the development of water projects, and limits the government's control on remote areas.

For the socioeconomic impacts, in the Kurdistan region of Iraq, the main occupations in the villages and remote areas are farming or shepherding, hence disabilities, such as the loss of a limb, caused by mine and UXO accidents effectively prevent the residents from carrying out their daily living activities, which in turn adversely affects their social and economic abilities. A person, who is an amputee, blinded or disabled is often no longer able to work as a farmer or a shepherd. This inability of disabled persons has a dramatic economic impact on the individual and their family's well-being. When a family head who supplies the main income has become a mine/UXO victim and been seriously injured, the dynamics of the entire family tends to change (Bier 2003).

In cases where the head of household is injured, the children more likely have to leave school to help the family to earn living expenses, or the mother who often stays at home with the children will need to work, also putting herself at risk of mines and UXOs. Hence, women and children with less experience and awareness of the danger of mines and UXOs have to fetch water and firewood daily, and as a result will more likely become victims themselves.

1.2 Food Security

Food security ensures that all individual at any time have access to physical and economic basic foodstuffs. According to the United Nations Human Development Report 1994, the availability of food is generally not a problem—the problem is its poor distribution and the lack of purchasing power of the population. In the past the problems of food security was addressed at national and global levels. However, their effects are limited. According to the United Nations human development report 1994, the key is to address problems related to access to employment or job opportunities, assets and revenues secured or security related economic factors.

Landmines and UXOs have tremendous impact on food security in regards to making arable land safe for cultivation. The mine clearance activity will ensure that agriculture lands return to productive use for cultivation and animal grazing. It enables the communities to become self-sufficient and less reliant on aid and public support, and enables vulnerable people to plant their crops, rebuild their homes and raise their families in a safe manner.

Regarding agriculture development, huge arable lands in the region are contaminated by mines and UXOs, and people have lived here for generations. In spite of their will and urgent need to cultivate the land, they cannot because of this, and are most likely forced to rely on exported food from bordering countries and on international aid, as fertile land lies unused. Hence, most of the contaminated village inhabitants cannot become self-sufficient. Moreover, these landmines also killed large amounts of livestock besides stalling the agriculture activities, reducing the production and food supply potential significantly.

The difficulties listed above, combined with the difficulties facing people in transporting food from other areas due to mined roads, will create a food shortage situation which results in food scarcity and malnutrition. Children are among those who suffer most from such a situation.

1.3 Health Security

Health security aims to ensure a minimum level of provision of basic health services and protection from disease and unhealthy lifestyles. In developing countries, the main causes of death were traditionally due to infectious and parasitic diseases, whereas in industrialized countries, the leading cause of death was from circulatory diseases. Lifestyle related chronic diseases are leading killers in all parts of the world. About 80% of deaths from chronic diseases occur in low and middle income countries. According to the United Nations report, in both developing and industrialized countries, threats to health security are usually greater for the poor people in rural areas and especially children than the urban and adult population counterparts. This is due to poor nutrition and inadequate access to basic health services, clean water and other basic necessities.

Mines and UXOs have enormous adverse effects on public health, beside the physical injuries that affect the inhabitants, it also has adverse effects on the Kurdistan regions' development process and outcome. The United Nations report shows that people most likely die due to poor water quality and sanitation, and disease and malnutrition than from direct injuries. The presence of landmines in the Kurdistan region hampers the wellbeing of the Kurdish society through its effects on water safety, agricultural development, campaigns of public health, and the socio-economic and emotional state of many Kurds residing in the contaminated communities throughout the region.

The landmines and UXOs have an impact on health security as they kill and maim innocent civilians long after the end of cross border conflicts, the blockage to water access causes malnutrition and waterborne diseases, mines on roads and pathways prevent the delivery of food and medication as well making it difficult for mobile health and vaccination teams to reach affected areas, in which diseases affecting children increase in frequency and effects.[2]

In addition, landmines may also affect villages that have lost one of their members due to a mine or UXO accident while fetching water from a fresh spring or a well. The inhabitants are subsequently more likely to avoid accessing the fresh water sources and to look for safer roads, but as a result access polluted water. Boiling this water in turn needs wood, leading the people to collect woods in the forest areas that are probably mined. This will hamper the people in contaminated areas from boiling their water and hence leads to infectious diseases (Kakar 1995).

For the campaigns of public health, the presence of landmines in the region's villages and communities have prevented the access and delivery of fundamental and critical health care to communities, as landmines are more likely to be found on roads that access public health clinics and public meeting places. Preventing the basic health care services from being offered will certainly affect the public health of mine-affected communities. In addition, the presence of mines and UXOs cause logistic and security problems for government health services, as well as for NGOs and international aid agencies active in the region in accessing communities in need of various healthcare programs and other forms of support such as a public health campaigns and mass immunization.

As illustrated above, the presence of mines and UXOs will prevent the public health activities in general and thus will lead to an increase in the risk of the affected population contracting infectious diseases. A number of preventable childhood diseases will also occur in areas suffering from coverage of low vaccination activities due to the presence of mines and UXOs (Krug et al. 1998).

1.4 Environmental Security

This type of security aims to protect people from the scourge of short term and long-term nature of man made threats in nature, and the deterioration of the natural environment. In developing countries, lack of access to clean water is one of the greatest environmental threats while in industrialized countries the main threat is air pollution. In addition global warming, caused by emissions of greenhouse gases, is another matter of environmental security which needs to be tackled.

Environmental security should aim to reduce or prevent the negative environmental consequences of war, environmental disasters at the level of war, erosion of the earth's carrying capacity, natural disasters and war and armed conflict resulting from environmental changes.

Landmines and UXOs have a huge negative impact on the environment. They cause land and soil degradation, habitat degradation, reduce vital resources, species losses, alternation of the natural food chain, loss of biodiversity, and serious limitations to agricultural production and productivity. It creates a huge distortion of economic development due to support system disruption of the biosphere's life and reducing the capacity of the environment for the supply of raw materials and natural resources. It is well known that landmines and UXOs block access to natural resources and increase the unsustainable utilization of marginal and ecological poor environments. It also destroy plants thus reducing biological diversity, contaminating the surrounding soil and water with high toxic substances and destroying the ecosystem through soil and water process disruption (Gangwar 2003).

Morin (1998) argued that the process of mine clearance in itself causes degradation and permanent damage to the environment. He further identified three problems that might be arisen in the process of mine clearance due to mismanagement of the environmental system. These are described as follows:

- Conservation of soil and erosion control: The different mine clearance methods by unsympathetic treating of the ground will cause different damage to the land, and improper demining methods will compromise soil integrity. The machine considered one of the most effective and quick ways of removing mines and UXOs repeatedly strikes the soil, creating shockwaves resulting from the grinding and chewing up of the ground by huge rotating drums, and as a result alters the sustainable soil structure and texture.
- Preservation of vegetation and natural wildlife: A process of random vegetation clearance as the first step of the physical demining process may affect the environment, and the vegetation area are more likely to be considered as food sources and natural habitats to different plants and animals that are placed in danger if these vegetation are cut or burned, and thus it threatens the ecological stability in the region for its very diverse inhabitants.

- Limits the adverse impacts of humanity: For the clearance process to start, it requires different types of human experts to be physically presented in the contaminated areas. The presence of humans in these areas requires different equipment and construction of different facilities. The effects of human presence on these lands that have been avoided for long time are severe. For example human waste, garbage, toxic oil, and fuel from vehicles used by these teams all harm the environment. These materials have severe impacts on local water supplies by way of natural runoff systems if they are improperly discarded.

In terms of water safety, in the region's remote areas contaminated by mines and UXOs, running water is not available and it is up to individual household heads to provide water for their family and for different purposes. Most of the mine- and UXO-related accidents are due to walking inside minefields while accessing water. Thus, landmines will obstruct the process of obtaining water in these remote areas. Hence, and by redirecting people from fresh water supplies, landmines put people at greater risk for physical injuries and contaminated water.

1.5 Personal Security

Personal security aims to protect people from physical violence from individuals and actors, violent states, domestic violence, or from adult predators. For many people, the biggest source of concern under such circumstances is crime, especially those classified as violent crime committed against them.

The landmines and UXOs destroy homes and prevent them from being reconstructed. It makes people risk their safety to generate income for their urgent survival and daily life. In addition, the consequences of being a landmine victim and becoming disabled will cause them psychological problems with undesirable consequences, which might further increase the degree of crime and violence in the communities.

1.6 Community Security

Community security aims to protect people from the loss of traditional values and relations of violence on sectarian and ethnic lines. The landmines and UXOs often threaten traditional communities, especially ethnic minorities. Nearly half of the countries in the world have witnessed some conflict between races. The United Nations declared 1993 the year of shedding light on the continued weakness of the three hundred million indigenous people in seventy countries facing a cycle of widening violence.

The impact of landmines and UXOs on communities can be considered an extension to the impact on individuals and families in the affected communities as well as on the livestock, arable land, access to clean water and reconstruction and rehabilitation programs.

Injuries from landmines will drastically impact the emotional well-being of Kurdish society. It has immense emotional and psychological effects, as a mine/UXO victim sees themself as useless or unimportant and this will affect the whole family situation and all members. Disabled victims more likely turn to alcohol and drugs, causing a serious public health problem in addition to injuries that make physical rehabilitation more difficult and ultimately affecting the structure and wellbeing of the entire societies.

1.7 Political Security

Security is political whether people are living in a society that respects their fundamental human rights or not. According to a survey conducted by Amnesty International, political repression and systematic torture was still practiced or enforced in 110 countries. Human rights violations are most common during periods of political turmoil (Elie, 2011). Along with the suppression of individuals and groups, governments may try to exert control over the ideas and information in the society. Since then, human security has received more attention from the major international development institutions, such as the World Bank.

Years of war and internal conflict have left Iraq in general and the Kurdistan region in particular littered with mines, UXO and stockpiled munitions. The hazards that people face, threatening normal activity and disrupting socio economic redevelopment, are enormous and have direct and indirect impacts, such as the socio-economic impact of mines and UXOs extending across accessing public areas, salvaging metal, accessing land for agriculture and grazing, and renovation and infrastructure. Indirect impacts include malnutrition as the arable minefields cannot be cultivated, starvation and the spread of infectious diseases, such as waterborne infections resulting from the difficulties in repairing water purification systems, and the provision of public health services in areas that are isolated due to mines and UXO contaminations (Kakar 1995). Thus, mines and UXOs in addition to taking lives, disable future generations, create insecurity and fear, and also hamper the return to normality.[3]

2. Summary of Human Security and the Political Impacts of Landmines

Landmines and UXOs cause a serious threat to the development and reconstruction of contaminated societies. They have strong negative effects on the process of societies' reintegration. This is in particular evident in communities that suffer a situation of political instability. The Kurdistan region was not excluded from the phenomenon of the presence of mines and UXOs and their deep and devastating effects. This is due to continuous conflict and wars in recent decades, the negative effects extending over time to include most of the communities in the region. The intensity of landmine and UXO prevalence differs from one community to another depending on their strategic locations.

Landmine and UXOs continue to threaten the lives of innocent people. Civilians are the most frequent victims in war (and in peacetime). Subsistence farmers, refugees, and villagers cultivating their fields, herding animals, collecting firewood, and bringing water are forced into dangerous mine-areas to access basic resources to satisfy their daily lives. Injuries caused by mines affect the lives of the victims and impact on their families and the medical infrastructure of the mine-affected country. Injuries often need a series of surgeries and treatments. The impact extends far beyond individual accidents and casualties. Huge areas of former battle fields have created acute social, economic and environmental harm through denying access to productive land areas for civilian use.

In sum, landmines and UXOs cause massive collateral damage to a society. They are different from all other weapon systems due to their indiscriminate nature and persistent threat after the termination of conflicts since they have the potential to harm long after the fighting has ceased. Lack of accountability, responsibility and public service provision often leads to continuous threat in rural areas. This chapter was organized around explanations of the multi-dimensionality of human security. The issues discussed included human security which include economic, food, health, environment, personal, community and political security aspects.

The concept of human rights in the third world covers mostly the political, civil, and economic aspects. The concept of human security focuses on the viewpoints of individuals to protect them from threats to their lives, livelihoods and dignity and to utilize their full potential. Human security which carries a viewpoint of individual centred security is essential to achieve national stability and at the regional and global levels by reducing political tensions.

The economic security aspect requires a basic guaranteed work income provided to individuals, or as a last resort of the safety net funded by the public sector. The economic impact of landmines and UXOs can be seen as one of the determinant factors of economic security as it prevents people in the affected area from working, and the survived mine victims with disabilities will face difficulties in finding employment without being dependent on their family, charity organizations or the state. In the Kurdistan region, the main occupations in the villages and remote areas are farming or shepherding, limiting the provision of suitable employment to disabled victims.

Food security ensures that individuals should have access to physical and economic basic foodstuffs. The landmine and UXO have tremendous impact on food security in regards to making the existing arable land safe for cultivation. The mine clearance has the objective of ensuring that agricultural land returns to productive use for cultivation and animal grazing. It enables the communities to become self-sufficient and less reliant on aid and public support, and enables vulnerable people to plant their crops, rebuild their homes and raise their families in a safe manner. In addition, they produce food supplies for the urban population and enhance food security and self-sufficiency.

Health security ensures a minimum level of provision of basic health services and protection from disease and unhealthy lifestyles. In developing countries, the main causes of death include infectious and parasitic diseases, while lifestyle related chronic diseases are leading killers in all parts of the world. The landmines and UXOs impact on health security as they kill and maim innocent civilians long after the end of conflicts, block access to water and cause malnutrition and waterborne diseases, and prevent delivery of food and medications and make it difficult for mobile health and vaccination teams to reach the affected areas by making roads inaccessible. The presence of mines and UXOs will prevent the public health activities in general and thus will increase the risk of the affected population in contracting infectious diseases.

Environmental security aims to protect people from threats in nature, and the deterioration of their natural environment. Lack of access to clean water is one of the greatest environmental threats. Environmental security should aim to reduce or prevent negative environmental consequences of war, erosion of the earth's carrying capacity, natural disasters, and conflict resulting in environmental changes. Landmines and UXOs cause land and soil degradation, habitat degradation, reduce vital resources, alter the natural food chain and create serious limitations to agricultural production and productivity. It creates huge distortions to economic development due to support system disruption and reduction in the capacity of the environment to supply raw materials and natural resources.

Personal security aims at protecting people from physical violence and crime exercised especially by the victims of landmines. Landmines and UXOs destroy homes and prevent them from being reconstructed. They cause people to risk their safety in generating income for their urgent survival and daily life. In addition, the consequences of being a landmine victim and becoming disabled will cause psychological problems with undesirable consequences that might further increase the degree of crime and violence in the suffering communities.

Community security aims to protect people from the loss of traditional values and relations of violence. Landmines and UXOs often threaten traditional communities, especially targeting ethnic minorities. The impact on communities can be considered an extended level to the impact on individuals and their families as well as to the livestock, arable land and access to clean water, and reconstruction and rehabilitation programs. In general, the injuries from landmines drastically impact the emotional well-being of Kurdish society.

Security is political whether people are living in a society that respects their fundamental human rights or not. Human rights violations are most common during periods of political turmoil. Along with the suppression of individuals and groups, governments may try to exert control over the ideas and information flows in society. Long-lasting periods of war and conflict have left Iraq in general and the Kurdistan region in particular littered with mines and stockpiled munitions. The hazards that people face which threaten normal activity and disrupt socio-economic development are enormous and are direct and indirect impacts of such inhuman threats.

Notes

[1] Human Development Report (HDR 1994), Published for the United Nations Development Program (UNDP), Oxford University Press.
[2] The World's Conference for Physics Therapy 2011, available at: (http://www.wcpt.org/node/32597), Accessed on July 22, 2011.
[3] See http://www.icbl.org/news/2003/343.php, Accessed on July 22, 2011.

CHAPTER TWELVE

MINE ACTION RESOURCES AND ITS PROGRESS

This chapter deals with the mine action's organization, capacity, responsibility, resources availability and funding, and the progress of its mission. After the transfer of power to the central government and agreement on a united but federal Iraq, the responsibility of mine action activities was also transferred from international organizations and local agencies to the central government. This is evidence of a lack of awareness and interest of the Kurdistan regional government to the mine issue, which exposes its population to future mine risks.

As previously mentioned, this action seems not to be appropriate considering the Iraqi government's crime against the Kurds through history in deploying huge amount of landmines in the region. Any collected information might be misused by the Iraqi army and security forces in a possible future conflict. Estimation of the mine cleaning capacity and speed and some kind of cost-benefit investigation can show how to increase this capacity by using international and national resources to build up local capacity, forcing the central government to allocate a special budget to clean up their remains and to rehabilitate the victims of their crime.

The Kurdistan region of Iraq is fatally contaminated by mines and UXOs due to different armed conflicts in different time periods. Landmines and UXOs have affected more than 12,800 people in the region since the early 1970s, of which more than 10,500 were killed. In 1991, the number of casualties rose to as many as 2,366 victims. The majority of these were civilians, of which 1,306 persons were killed. In addition to risks of accidents and fatalities, landmines constitute a substantial barrier to rehabilitation and development activities in many areas in the region.

The resources available to mine action need to be accurately managed in order to respond to the regional humanitarian and development priorities in an effective manner. The humanitarian mine action in the Kurdistan region of Iraq needs to play an imperative role in supporting and enhancing the socio-economic development here. To achieve this goal, information about socio-economic benefits needs to be accurately collected, assessed and periodically updated to enable parties to conduct different cost-benefit analysis of mine clearance activities.

The Kurdistan region witnessed a drastic economic growth since the liberation of Iraq in 2003. As a result of the changed business environment, a huge construction boom has started throughout the entire region. However, the presence of mines and UXOs will hinder the sustainable growth in the remote and contaminated areas. Livelihood systems are not secured, in particular for most rural communities which comprise predominantly subsistence farmers. During 1992, the invasion of the Iraqi regime into the Kurdish territory meant that hundreds of thousands of people were displaced and moved to the Iranian and Turkish borders. This is the main argument against the transfer of responsibility for mine action activities from local to central government and agencies.

The extent of the mine and UXO contaminated areas in the Kurdistan region is disputed and those parts of it should be given priority for clearance. By disputed we mean that the extent of contamination is significantly underestimated. Changes in the demography and rapid urbanization are evidence of such underestimation—for example, the depopulation of rural areas helps in not discovering many of the contaminated areas and therefore it is strongly recommended to revise these estimates.

Beside the reduction of accidents through mine clearance, it is noted that other measures such as mine awareness education programs need to be strengthened in order to further reduce the number of casualties. There is evidence of strong positive effects from such programs which in turn put more pressure on the agencies to allocate the necessary resources and undertake minefield clearance.

The rest of this chapter is organized as follows: a brief review and discussion of the need for conducting cost-benefit analysis of mine action activities is discussed in section 1, the strategic vision of mine action in section 2, mine action organization, resources, responsibilities and different stakeholders are discussed in section 3 and 4, and finally a summary and the concluding remarks are discussed in section 5.

1. The Need to Conduct a Cost-Benefit Analysis

In the assessment of the mine action achievement aiming to reduce human casualties, one should account for the potential contribution of the Kurdish landmine victims in the production activities in the region. However, this assessment should not be considered a substitute for the value of human life, instead, saving life, reducing pain, and reducing casualties risk should be given the highest priority.

The clearance activities of mine and UXO contaminated fertile lands in the Kurdistan region is to hand over these arable agriculture lands to their farmers and owners. The main food production in the region is grain, hence the output from crops need to be used as a guideline for the production value as well as livestock as a complementary trade. Clearing the water supply will be useful in bringing life to these areas to enhance cultivation and raise livestock. These are indications that landmine damage to domestic animals might constitute a major economic threat as it will take years to build up the stock of animals again.

Transportation is vital for marketing agricultural products and access to health care, education and production inputs. For the provision of more convenient transportation options, contaminated roads need to be cleared which will then provide more options to travel and reduce travel time and travel cost. Moreover, clearing schools, water paths and water wells, health care centres, religious sites and tourism areas will also provide benefit for the users of the cleared roads to access alternative facilities in other distanced areas.

As mentioned by Byrd and Gildestad (2001), the impact of landmine contamination and related accidents in general can be categorized in terms of socio-economic behaviour into three types. First, the impact of landmine accidents on human life resulting in loss of life or amputation or paralysis, which in turn will have negative effects on human production and welfare. Second, landmines block access to necessary resources for human production and consumption. Third, the presence of landmines cause a distortion of human attitude for various reasons, such as the socio-economic harm due to longer-distance travelling as shorter roads are blocked by mines and UXOs.

A high cost will be incurred as a result of dealing with each problem in the different criteria mentioned above associated with different mine action resources used and activities such as clearance (manual, mechanical, dogs, etc.). Among other important and costly activities are those related to demining accidents during a mine operation, provide mine awareness educational programs and the various field-support services during the operation.

Moreover, the incurred cost of landmines can be considered from a socio-economic impact perspective, which are mainly outcomes such as death or injuries of people, death of livestock, blocked access to infrastructure such as water supply, roads, health centres and schools, and blocked access to religious sites and tourism. The cost is attributed to the management of risk and to the loss of potential revenues from production and tourism, as well as those not measurable like benefits from visiting religious sites.

For example, when there is a mine threat from accessing the contaminated water supply locations, the burden can be weighed against increased travel time and cost caused by fetching water from distant water wells, and increased waiting time in the event of insufficient supply to meet demand. Delays in the supply of water are important to, for instance, construction and irrigation, causing costs in the form of direct labour and indirectly in the form of loss of crop yields and penalties due to violation of construction contracts with deadlines.

Hence, accordingly, a mine action project's cost-benefit analysis can be calculated as a net benefit (benefit minus cost) and divided by cost to obtain a benefit/cost ratio (BCR), to be used in project rankings. Therefore, the higher the value of BCR will be associated with the value of the project undertaken. A negative BCR indicates that the project cannot be assessed based on its socio-economic indicators and needs further investigation, referring to issues such as the future expected rate of reducing landmine accidents needing consideration to fully undertake different aspects of certain mine action projects.

A cost-benefit analysis involves four steps. First is the need to identify all the relevant costs and benefits. Second, it is important to place monetary values on these costs and benefits where possible. Third, an attempt is made to convert the stream of future costs and benefits into comparable terms known as discounting to present value. Finally, one should calculates measures by which a project or program may be judged.[1] Here, reference is made to the net benefits from alternative projects.

A cost-benefit analysis typically involves an investment and a stream of net returns that follow over a number of years, for example the value of increased agricultural output minus the mine clearance and production costs. However, the study of demining does not follow this usual path as landmine clearance (the investment) is more likely to take place over a long period of time, and the benefits are the saved costs resulting from having fewer mines over time as well as other benefits related to life and safety of the population.

Landmine clearance has numerous benefits in which some are of measurable or monetary quantitatively while others are not. At least three main benefits of mine action are observed from three aspects. First, saving of lives and prevention of disabilities, both of which can be valued by applying the average number of years by which a productive life is shortened to the value of potential output lost in each year. Second, medical costs that include the short-term cost of medical treatment for survivors and those ongoing of amputees for artificial limbs. A landmine injured victim, based on estimation of the international committee of the red cross (ICRC), will probably stays in the hospital for an average of three weeks and an amputee stays for five weeks as they require four operations, on average. They also need antibiotics and blood transfusions. Moreover, amputees should have prostheses replaced every three to five years for adults and every six months for children. Third, saved agricultural output from land that otherwise would not be used or would be underutilized as a result of mines, assessed by applying value added per hectare (Ghali 1994).

In general, the mine action activities, regardless of location and conditions, can be seen as beneficial in comparison to their incurred costs. However, it is somehow difficult to justify this quantitatively as the benefits are associated with many other immeasurable factors. The mine action activities allow the beneficiaries to access resources and infrastructures of value for personal and societal. For example, access to education and health are important for the well-being and development of a country's economy and population. A country should be able to use its available resources most efficiently such as water and power supplies, and it should have access to useable infrastructures such as roads. However, it should be noted that the type of road, power supply, water well, or school that is most worth accessing will vary from case to case. Rural areas are often neglected in resource allocation and under such conditions the limited resources should be available for use. Hence, prioritization and accurate analysis and planning are imperative before starting mine action activities.[2]

2. The Strategic Vision of Mine Action

Mine action in Kurdistan region has been a blessing to the population and the economy of the region. The outcome can however be improved following more resources, better techniques and taking advantage of existing experience locally and internationally. Like general development vision by the regional government, the mine action suffers from lack of clear vision and strategy. It needs to make a clear feasibility study in the Kurdistan region to show long-term vision, which it should have the following targets:

- Reduction of mine and UXO accidents and casualties as a main target.

- Clearance of the contaminated path roads in order to easily provide access to fundamental services such as water sources, roads, health centres, schools and markets for agricultural inputs and outputs.
- Agriculture lands to be free from landmine contamination, which effectively hampers economic growth in rural areas.
- Continually demarcation of the non-cleared landmine areas and increase in awareness through training and education rate.
- Regular update of clearance prioritization strategy based on socio-economic needs at the local level.
- Assist the surviving landmine victims through a better health care provision, rehabilitation, prosthetic limbs centres and vocational training.

3. Mine Action Organization and Stakeholders

The national regulatory authority for the sector of Mines and UXOs in the Kurdistan region of Iraq is a governmental institution responsible for the regulation and coordination of all mine related operators who work on reducing the impact of unexploded bombs, ERW, artillery shells, landmines and UXOs. The national authority counterpart is called the Iraqi Kurdistan Mine Agency (IKMAA). The overall vision of IKMAA is to enable the people in the region to live free from the threat of landmines and UXOs, help to promote its development, and to completely integrate the landmine and UXO victims into Kurdish society. Despite such noble goals, the IKMAA has not achieved much since transfer of the authority from local to central agency. In addition to lack of performance, the information available on landmines in the region can be misused by the Iraqi army and security forces in case a conflict erupts between the regional and central governments.

IKMAA has been established as a national authority after the coalition forces' invasion of Iraq in 2003 and the evacuation of all UN agencies operating there in general, and in the Kurdistan region specifically. Despite this, a strong network of local government, international support, and operator agencies still continues to deliver key services in the focus areas of clearance, risk education, victim assistance and increasingly in quality control, information management and capacity building through vocational training. However, a transfer of responsibility from regional to central authority might have significantly reduced the effectiveness of these organizations in decision making about different activities and preferences.

4. Resources for Mine Action

According to Trevelyan (2000), mine action resources can be provided in the form of, for instance, equipment, technical assistance, training, managerial assistance, resource mobilization assistance and financing. However, the most commonly provided resources are international aid funds, in-kind support from international aid donors, and direct and indirect host government support and funding. In addition, the use of military personnel in demining operations, other wealthy donor governments, the United Nations and other international organizations, and benefactors and philanthropists are among other important resources.

The IKMAA have different financial sources such as resources allocated by the Kurdistan Regional Government (KRG), resources provided by UN developing program UNDP, and donor country contribution. The utilization of all resources provided for IKMAA are closely coordinated with concerned national or regional action authorities.[3] Despite high oil, gas and tariff revenues by the regional government, local resource allocation is insufficient to correspond to the mine action activities needed. Without external support, the results will be poor and inadequate. Very likely, the flow of external resources will be reduced in the near future due to their priorities for countries with much more limited resources than the Kurdistan region.

The development in the region is characterized by being business driven and not based on a proper resource allocation and cost-benefit analysis from net social value returns perspective. Lack of interest in mine action is due to no investment and profit opportunities for private business. The government gives the highest priority to the construction of different public development infrastructure projects. These projects are performed by corporations which are often owned solely or jointly by the decision makers. The agricultural sector has been neglected as part of the reconstruction and development programs, and the import and distribution of consumer goods have been more profitable and less risky business activities than local production and long term investment.

According to the UN electronic mine information network in 2011, different international agencies undertake different mine action aspects in the Kurdistan region, which are classified as follows.[4]

4.1 The Community Based Mine Risk Education

The community-based mine risk education (CBMRE) program is run and funded by Handicap International (HI). The program offers MRE to the higher priority and contaminated villages in the three governorates of the Kurdistan region. CBMRE activities can be summarized as follows:

- Select the most at risk communities and introduce the project's concept and methodology.
- Train community focal points on CBMRE.
- Support the established focal points to implement community mapping and seasonal calendar.
- Develop referral systems between communities and stakeholders and disseminate the information.
- Deliver mine risk education sessions at community level.
- Develop training on disability and its better management at community level
- Conduct disability awareness sessions.
- Define strategy, priorities and design project.
- Conduct training of local partners on CBMRE and monitor activity.
- Identify and disseminate lessons learned and methodologies.

4.2 The Conflict Recovery Program

The conflict recovery program is funded by different sources and run by the MAG in cooperation with the IKMAA, the General Directorate of Mine Action (GDMA), the Kurdistan regional government ministry of education, and several other national non-governmental partners.

The program aims to save lives and to improve the safety and futures of people affected by years of conflict and war and its multifaceted remnants. This can be achieved by clearing contamination areas of highest impact.

The conflict recovery program aims at the following activities:

- Conduct demining clearance and destruction, demarcation, and technical surveys.
- Prioritize clearance and destruction activities, and conduct pre- and post-clearance assessments impact of operations.
- Use mine detection dog teams and mechanical assets to increase the efficiency of manual mine clearance.
- Conduct mine and small arms' risk education MRE to targeted vulnerable groups in the society, including internally displaced persons (IDPs), nomads, and children.
- Deliver capacity building training to support development of local and sustainable MRE capacity.

- Research and development of mechanical clearance assets.

4.3 Technical Support and Landmines Impact Survey Completion

Technical support and landmine impact survey completion program is multi source funded and run by the Information Management of Mine Action Program (iMMAP). The program aims to continue supporting humanitarian mine action for coordination and capacity building through proving technical support to the Information Management System for Mine Action Database (IMSMA), expansion of Operational Activity Security Information System (OASIS) distribution among all humanitarian mine action stakeholders, and support landmine impact survey activities.

4.4 The Victim Assistance Capacity Development Program

The victim assistance capacity development program is run by iMMAP and aims to train rehabilitation technicians through adapting course curriculums that will best enhance the prosthetics and orthotic skills of the technicians. In addition, it provides the prosthetic limb centres in the region with the necessary materials and components to execute the training courses. The program will build a cohesive relationship with the ministries of education and health in the Kurdistan region to create awareness of rehabilitation needs, to increase the level of education available to technicians, and to improve rehabilitation capacity in the region allowing the victims to return to rewarding and productive lives.

4.5 MRE for Reducing Injuries and Casualties, and Victim Assistance

The mine risk education for reducing injuries and casualties program and the victim assistance program is run and funded by the UN Children's Fund (UNICEF) to strengthen capacities of national mine action decision makers, trainers, teachers, health workers, boys, girls and community members. In addition, the two programs equip these groups with training, materials, and other necessities to reduce the number of landmine victims and to support national efforts to establish a mechanism for victim surveillance, victim assistance, and reintegration of victims and survivors into their societies.

The main activities of the MRE for reducing injuries and casualties and victim assistance program are as follows:

- National capacity development for planning, managing, implementing, monitoring and evaluation of mine risk education.
- Mine risk education and public information campaigns on risk and threats of landmines at risk populations for prevention and reduction of injuries.
- Mine risk education needs assessment in contaminated areas.
- Victim surveillance and assistance and promotion of survivor's rights in cooperation with other actors.

Institutions and organizations involved in mine action activities have an important role in reducing the impact of landmines and contributing to the solutions of the problems caused by mines and UXOs in the contaminated communities. They play a key role of speaking on behalf of the weakest in the society and in bringing their well-being and mine issues to the decision makers' attention. However, there are still opportunities for improvement in the execution, coordination, efficiency, effectiveness, innovation and proper utilization of resources as well as promoting competition in mine action activities. Overlapping of work should be avoided to increase effectiveness and to save resources from being wasted (see Steven et al. 1996).

The landmine action organization believes that local structures with appropriate support are best placed to promote community based responses to the problems caused by conventional weapons in society and especially mines and UXOs.[5] Local agencies can work in partnerships and networks of like-minded NGOs, states, non-states actors, community based organizations, donors and others to achieve the goal of landmine-free environment. Innovation should be at the centre of landmine action thinking through using the project work to influence the political and legal agenda to promote new ways of working.

The landmine action's values determine local agencies relationships with all stakeholders including beneficiary communities, network organizations, donors and employees. Transparency, accuracy and honesty will be crucial to the credibility of landmine action's work and the key to partnerships with states, civil society groups and community based organizations. A commitment to effectiveness and cost efficiency will drive the operational activities and inform the fundraising.

5. Summary of Mine Action Resources and Progress

This chapter deals with the mine actions organization, capacity, responsibility, resource availability and funding and progress in its mission. In recent years, the mine action organization has witnessed major changes. After the transfer of power to the central government and agreement on a united but federal Iraq, the responsibility was transferred from the international organizations and local agencies to the central government.

From the Kurds' perspective, the above-mentioned transfer of responsibility to the irresponsible central mine action agency to plan and implement is a major step backwards. This is an evidence of lack of awareness and interest of the Kurdistan regional government to the mine issue. This action seems not to be appropriate considering the Iraqi government's crime against Kurds, including the systematic laying of mines in the Kurdistan region. The collected information might be misused by the Iraqi army and security forces in a possible future conflict. Regional government should promote local capacity and force the central government to allocate a special budget to clean up their remains and to rehabilitate the victims of their crime.

The Kurdistan region of Iraq is fatally contaminated by mines and UXOs due to different armed conflicts in different time periods. Landmines and UXOs have created more than 12,800 victims in the region since the early 1970s, of which more than 10,500 were killed. Landmines constitute a substantial barrier to rehabilitation and development activities in many areas of the region. The resources available to mine action need to be accurately managed in order to respond to the regional humanitarian and development priorities in an effective manner. The humanitarian mine action in the region needs to play an imperative role in supporting and enhancing its socio-economic development.

The extent of the mine and UXO contaminated areas are disputed and especially those parts that should be given priority for clearance. By disputed we mean that the extent of contamination is significantly underestimated. The changed demography and rapid urbanization is evidence of such underestimation. The depopulation of rural areas has led to the fact that not all contaminated areas have been discovered. It is strongly recommended that the estimates of contaminated areas are revised.

This chapter discussed the need for conducting cost-benefit analysis for mine action activities, the strategic vision of mine action, mine action organization, resources, responsibilities and different stakeholders.

The impacts of landmine contamination and related accidents in general can be categorized in terms of socio-economic behaviour into three types: (i) impact of landmine accidents on human life causing death or amputation or paralysis and reducing production, (ii) landmines block access to necessary resources for human production and consumption, and

(iii) the presence of landmines causes a distortion of human attitude for various reasons, such as the socio-economic harm from longer-distance travelling.

There is a need to conduct cost-benefit analysis. A cost-benefit analysis involves four steps: (i) identifying all the relevant costs and benefits, (ii) places monetary values on these costs and benefits where possible, (iii) attempt to convert the stream of future costs and benefits into comparable terms known as discounting to present value, and (iv) calculate measures such as net benefit by which a project or program may be judged. Landmine clearance has numerous benefits, some measurable quantitatively or monetarily, some not. At least three main benefits of mine action are observed: (i) saved lives, disabilities, and productive forces, (ii) medical costs that include the short term cost of medical treatment for survivors, and (iii) saved agricultural output from land that otherwise would not be used or would be underutilized as a result of mines contamination.

The mine action in the Kurdistan region has been a blessing to the population and the economy of the region. The outcome can however be improved following more resources, better techniques and taking advantage of experience locally and internationally. Like the general development vision of the regional government, mine action suffers from the lack of clear vision and strategy. Mine action in the Kurdistan region needs to make a clear feasibility study to show the long-run vision in which it should have important targets.

The Iraqi Kurdistan Mine Action Agency (IKMAA) has different financial sources including the regional government and international organization. However, the local resource allocation is insufficient to correspond to the mine action needs. Therefore, without external support the results will be poor and inadequate. Very likely, the flow of external resources will be reduced in the near future. As part of a business driven development strategy, the regional government has so far given the highest priority to construction of different public development infrastructure projects. The agricultural sector has been neglected as part of the reconstruction and development programs in the region. The import and distribution of consumer goods has been a more profitable and less risky business activity than local production and long-term investment.

Notes

---

[1] Brealey, R.A., Stewart C.M. (1991), *Principles of Corporate Finance*, 4th ed. McGraw-Hill.

[2] Geneva International Center for Humanitarian Demining (GICHD). *A guide to Mine action*. Switzerland, 2004.

[3] According to Landmine Monitor Report 2005, the National Mine Action Authority has stated that it received some $61 million in donations in 2004. Landmine Monitor has been able to identify $58.7 million in contributions to mine action in Iraq in 2004 from thirteen donors. This includes donations to NGOs and international agencies operating in Iraq, and some in-kind contributions. The Landmine Monitor has identified $55 million in international contributions to mine action in Iraq from fifteen donors in 2003.

[4] Electronic Mine Action Information, available at: http://www.mineaction.org/projects.asp?c=14&pg=2

[5] See http://www.landmineaction.org/, Accessed on July 26, 2011.

CHAPTER THIRTEEN

MINE RISK EDUCATION AND OTHER PREVENTIVE MEASURES

Mine action refers to those activities that aim to reduce the social, economic, and environmental impacts of landmines and UXOs. Mine action is not just about demining, it is also about people and societies and how they are affected by landmine contamination and their well-being. The objective of mine action is thus to reduce the risk of landmines to a level where people can live safely and under conditions such that economic, social, and health development can occur free from the constraints imposed by landmine contamination, and in which the victims' different needs can be addressed effectively.

As illustrated in previous chapters, mine and UXO contaminated areas will not probably be cleared in the near future. Therefore, the mine risk education program is an essential prevention program in reducing the number of deaths and injuries caused by mines and UXO. The mine risk education program aims to teach local populations in contaminated areas to avoid entry to the suspected minefields. However, it may not be easy to avoid entrance to certain areas as they are vital for a living, but it can be considered as an effective preventive measure to effectively reduce the landmine accident rate.

Mines and UXOs continue to be considered a mass international problem with long-term individual, social and economic consequences. In earlier chapters we discussed the multitude and correlated negative effects of landmines on rural lives and their well-being. In this regards we specially emphasized the effects on health care program and agricultural production situation, as well as normal human rights in mobility and its negative effects on families' mental and physical health.

Programs and campaigns based on sharing information, teaching and identifying the methods of avoiding mine risks and protecting people from being victims are all considered among the major components of a standard mine risk education program employed by national and international mine action organizations.

Mine risk education is part of a global assistance approach even if, in many cases, it is carried out in coordination with demining operations as a complete and comprehensive program. Such a program focuses on accident prevention and targets the population at risk, and has the ultimate objective of bringing life back to contaminated regions and reducing population flows, migration and poverty.

According to the GICHD report (2010),[1] a vast amount of money and other valuable productive resources are spent on an annual basis for mine action programs which reduce the effects of landmines on innocent people. Even though these programs employ thousands of people for this purpose, there is a lack in exchange of information such as lessons learned through relevant practical experience. Accounting for and management of feedback seem to be important in the evaluation of resource utilization and necessary reallocations in activities with the highest pay off in reducing risks.

Different kinds of resources for mine education have been used throughout the process of mine action such as media which is represented by posters and pamphlets. However, these types of media do not have a long lifespan as they are text dependent and might not be effective and suited in multi-language or low-literacy communities. Thus, in the design and planning of the media component of the action, as carriers of information an account should be made for the multidimensional aspects of the mine problem.

Another type of media used is the commonly used channels of radio to reach most remote areas. Among other channels are video and educational TV programs aimed at accessing populations in main urban areas and cities at different age groups. It is worth mentioning that the absence of electrification, coverage and necessary equipment might be a hindrance to using this technology more effectively in accessing the population in rural and remote areas. The national grid system is still under construction and expansion to cover such remote areas.

In general, the mine education programs are used to study the ways in which fatalities can be prevented or reduced in respect to their different impacts. In design and provision of programs, they must target different age groups with different abilities to understand the message. In particular, attention should be paid to the groups of the population with higher risk of being victims of landmines. Some groups are aware of the risk but are compelled to take risks due to the lack of alternatives of obtaining necessities such as water and energy in different and safe ways.

Given the results from chapter twelve on resources, in this chapter we discuss various educational and training programs, with the objective of building the necessary capacity for demining and also preventing accident rate and reducing their negative impacts. In Section 2 we discuss the mine awareness program cycle including the feasibility study, requirement assessment, program planning, monitoring and evaluation. Section 3 contains discussions of the communication and mine risk education with focus on interpersonal, small and traditional media and mass media, community liaison channels, as well as mine risk education strategy for the community. The final section provides a detailed summary and conclusion of this chapter.

1. Mine Awareness Program Cycle

According to the mine action standard, four steps of the program cycle are required to design and provide an efficient mine awareness education system. The different and interrelated consecutive steps are briefly described as follows.

1.1 Feasibility Study

The feasibility study helps to identify the needs for the mine awareness program in specific regions. It determines the mechanism of conducting the program in a way to integrate with other mine action activities, or to be treated as an independent program run and funded by a particular organization, often an NGO. The feasibility study aims at offering a preliminary study of the level of landmine threat in terms of area size, location, its potential and actual impacts on the affected community, and the degree in which the existing mine action activity and their outcomes is undertaken in the region.

The feasibility study also aims to identify the needs for further assistance in this regards. The feasibility study should be initiated firstly through conducting different levels of interviews with key informants in the affected communities such as military, religious, and political personnel, and with relevant government authorities, community leaders, involved mine action agencies including UN and different NGOs. Moreover, the study should also consider additional sources of information involving official records such as population statistics, accident and incident data, and other relevant documents such as media reports, related research reports, and books published.

The feasibility study should also make a capacity assessment for the current mine awareness program and whether it has necessary resources and capabilities to undertake and maintain such program. In particular, the feasibility study aims at assessing capacity through addressing a number of issues. These include whether the current program has enough resources and knowledge experience, if there is sufficient manpower at the field level to efficiently run the program, whether the affected areas are politically permitted to be accessed, the length of time for sustainable financial support, what the operational constraints that hinder the effectiveness of the program are, and what the additional needs are and in which area the extra support should be utilized to address those needs with the highest priority.

Moreover, based on the criteria mentioned above, the feasibility study will enable a decision as to whether there is a need to undertake a mine education program or not, and whether resources available have alternative uses and must be used for purposes that generate the highest social returns.

1.2 The Required Assessment

The process of identifying the required assessment is about collecting and analyzing information and relevant data in a systematic way to accurately identify the at population at risk, its size in terms of the level of the risk, and to identify the most appropriate strategy for the mine awareness program. This aims to conduct a successful development of all phases of the mine awareness program. Moreover, the level of the needs assessment will be determined by available financial and various available logistical resources for future and potential program implementation.

The process of data collection requires intensive training on proper management of the needs assessment, resource concentration, sufficiency in resources availability, and accuracy and commitment. This information is necessary in order to generate a population representative and reliable dataset for further analysis to be conducted for different purposes in the overall mine awareness program strategy.

The data collection should address the study of the effects of landmines on communities and the level of the response by different mine action aspects. This might include data on fatality rates, the level of blockage to important infrastructure and facilities, blocked access to health centres, the socio-economic impacts of landmines, and the magnitude and the geographical concentration of the problem. Further issues also need to be addressed in the assessment such as the drivers of the risk taking behaviour, the suitable method for learning and communicating, and demographic and other data related to landmine victims are also essential in this context.

Furthermore, for an effective implementation of the mine awareness program, needs assessment should in parallel also gather technical information about the mine and UXO types, locations and contaminated land type, the roles of individual family members in the affected communities, the literacy rate and the level of education, the political and historical background of the landmine problem and the affected communities, the mine and UXO affected country's political situation to assess the degree of humanitarian aid, means of communication such as languages, dialects, traditional media, and material, and familiar communication methods to the communities under study. Communication is vital in the process of mine awareness program as it can be used to inform people at risk, to establish self-behaviour in dealing with the risk of landmines through teaching mine safety skills, and it can be further used in the socio-legal environment to support different mine awareness activities.

The needs assessment, by addressing the above mentioned issues, will have the necessary information and resource requirements to design the four necessary and interrelated landmine surveys to be conducted including the general or level one survey, the landmine impact survey, the technical survey, and the completion survey, and hence to conduct them in a chronological manner.

The analysis of data collected by needs assessment in qualitative and quantities methods form should be based on categorizing the level of high risk behaviour into three categories: (i) people who are aware of the existence of landmine problems who are able to minimize the risks, but who continue high risk behaviour, in which case attitude, social problems and economic needs are the factors that need to be considered; (ii) people who have no knowledge about the safest behaviour when exposed to landmine threats, and are aware of the dangerous but have no necessary knowledge to reduce the risk, and finally; (iii) people who are not aware of landmines and have no knowledge about the existence of the landmines and the degree of risk.

1.3 Program Planning

The most efficient methods to address needs should be identified through appropriate program planning in which it identifies the overall goals and specific objectives, identifies the minimum acceptable standard and the activities that meet the standard, and establishes methods and structure for evaluation and post impact assessment. The program planning should aim at assessing and planning the philosophy of the agency in which it implements the program, the terms of the donors, the government body strategy, the degree by which it identifies the needs of the contaminated community, and the strategy of mine action and other relevant humanitarian agencies.

It is worth mentioning that without well thought out program planning, it is very unlikely that scarce resources are coordinated and used most effectively, resulting in the objective not being achieved. Often local capacity is limited in this area, therefore skilled and experienced labour from humanitarian mine agencies are the main contributors to the effective and comprehensive program planning.

1.4 Monitoring and Evaluation

Monitoring and evaluation are important tools for providing feedback and information on conducted activities of mine risk education. They aim at improving the mine risk education program over both short and long run so as to make it more reliable, strengthen the learning objectives, and strengthen the support mechanisms to affect any possible broad and necessary changes in policy. It helps to identify the limitations and strengths to guide a better allocation of resources.

Monitoring system should be developed to cover progress implementation and outcome assessment for mine risk education programs. It should aim at comparing the effects of the conducted program on landmine-affected populations. These effects can be categorized into the three types described below.

The first type involves the comparison between the rate of landmine fatalities before and after conducting the program. However it is important to take into consideration other factors that might influence the rate of accidents, including refugee and internally displaced persons' movements from one community to another, security initiatives and the cultivating and harvesting seasons in agricultural lands that might increase the rate of fatalities.

The second type concerns the general knowledge and confidence identified during the process of mine awareness phases, whether the community's level of confidence facing landmines and the knowledge about the existence of the threat is improved or not.

The third type elaborates blocked access to land, making the people more aware of identifying cleared land from contaminated land to promote resettlement activities.

2. Communication and Mine Risk Education

In order for any mine risk education program to be effective, it should use different kinds of communication processes such as reading, writing, dialogues, questions and answers, TV, and learning in the classroom. In this respect, communication techniques such as using the voice, facial expressions, movement, and communication media play a major role. These can be categorized into different types as follows:

2.1 Interpersonal Communication

The person-to-person or interpersonal communication includes direct contact which involves debates and questions and answers that facilitate the joint understandings of the situation. Moreover, family conversation, discussions with health service professionals, religious and community leaders, teachers, social organizations, and government officials, and child to child communication are other important and complementary interpersonal communication channels which are used differently by different groups, and with heterogeneous impacts depending on the individual responsiveness.

2.2 Small Media

Small media are tools such as leaflets, posters, records or cassettes, broachers, video materials, charts, T-shirts, and others items which are used to support the larger communication tools and interpersonal communications. Their preparation and distribution lead to the enhanced formation and activity of different communities. It is crucial for the mobilization of available resources to effectuate the mine risk education provision, participation, its effects and coverage.

2.3 Traditional Media

Art performance used to convey information in an entertaining way, especially live performance such as drama, puppet show, theatre, and storytelling and dances, is a means to provide opportunities for interaction between performers and the audience who are the target for conveying information. Traditional media is often a good complement to the small media that helps to increase the effect of such activities. They often have other impacts than information in the form of rehabilitation and in particular preventing isolation and a better participation in community activities.

2.4 Mass Media

An indirect way of communication, such as newspapers, magazine, books, radio and TV, provide a convenient way to reach the target population without planning and implementation of some form of interpersonal communication. However, with the lack of social infrastructure and people with a high level of education, the cost and difficulties of delivering mine risk education through mass media communications have rapidly raised as the media needs appropriate facilities to deliver the necessary information in an cost effective and high impact way.

2.5 Community Liaison

The process of exchange of information between national authorities and mine action agencies is useful for the communities as it makes them informed of the conduct of any mine action activity and its type and duration. In addition, it makes the people informed of the details of the activities that have taken place such, as demarcation or clearance activities.

The major obstacle to good planning, to prioritize the clearance, and to implement the mine education program at the community level is the lack of data. Therefore, an appropriate mine awareness program will gather as much reliable information as needed which must be exchanged with the other mine action components of survey, clearance and victim assistance. Different mechanisms may be used for data collection at the community level such as interviews and discussion groups. It is vital for any interviewing process which results in the creation of data to identify how to make it useful for the interviewed and for the community which must be made interested in the process. The collected information for the mine risk education context is needed to evaluate the achieved works in terms of making necessary changes and additions to the current program and to develop a capacity of community liaison.

Moreover, different individuals such as the local military and police, teachers, local politicians, religious leaders, community elders, the women's community, health officials, and business people within the community should be approached. These individuals and personalities will give relatively different answers, and individuals with key functions should be encouraged to become actively involved in the process of collecting data and operation of the mine awareness program.

2.6 The Mine Risk Education Communication Strategy

It is essential to have a communication strategy which is based on the communication processes, techniques and channels that are most suited to specific audiences. The strategy should be aimed at providing information, assurance, support, and social atmosphere needed to encourage a mine-safe attitude. It should also be aimed at teaching the needed skills and to persuade people to share information.

3. Summary and Conclusion on MRE and Preventive Measures

Mine action refers to a number of activities aimed to reduce the social, economic and environmental impacts of landmines and unexploded ordnances. It is not just about demining but also about people and societies, and how their well-being is affected by landmine contamination. The objective of mine action is thus to reduce the landmine risk to such a level that people can live safely from the constraints imposed by landmines.

The mine risk education program is the essential means of reducing the number of deaths and injuries caused by mines and UXO. It aims to teach local populations in contaminated areas to avoid entry to the affected or suspected areas. It may not effectively prevent entrance to risky areas, but it can be considered as an effective preventive measure to effectively reduce the landmine accident rate. Here the emphasis is on the reduction of the effects of landmines on health care and the agricultural production situation, as well as normal human rights in mobility and its negative effects on families' mental and physical health situations.

Different programs and campaigns based on sharing information, teaching and identifying the methods of avoiding mine risks, and protecting people from becoming victims are all considered among the major components of a standard mine risk education program employed by national and international mine action organizations involved in mine clearing and accident prevention activities.

Different kinds of resources for education have been used throughout the process of mine action such as posters and pamphlets. Although these types of media have a short lifespan as they are text dependent and may not be effective and suited to multi-language or in low literacy communities. Thus, the design and planning of the media component of the action should be made by accounting for the multidimensional aspect of the mine problem. Another type of media used is radio which reaches most remote areas. Among other channels are video and educational TV programs aimed at accessing the populations in the main urban areas and cities at different age groups. The use of several such channels is restricted by accessibility of electricity distribution in remote areas.

In this chapter we have discussed various educational and training programs with the objective of building up necessary capacity to demining and to lower the accident rate and reduce the negative impacts. The mine awareness program cycle was introduced including a feasibility study, requirement assessment, program planning, monitoring and evaluation. These steps of program cycle are required in order to design and provide an efficient mine awareness education. The different and interrelated consecutive steps are briefly described as follows.

The feasibility study helps to identify the needs for mine awareness programs in specific regions. It determines the mechanism of conducting the program to be integrated with other mine action activities or to be treated as an independent program run and funded by particular organizations, often an NGO. It offers a preliminary study of the level of landmine threat in terms of area size, location, its potential and actual impacts on the affected community, and the degree in which the existing mine action activity and their outcomes are undertaken in the region. It also aims to identify the needs for further assistance.

Requirement assessment involves collecting and analyzing information and relevant data in a systematic way to accurately identify the population at risk, its size in terms of the level of risk and to identify the most appropriate strategy for a mine awareness program. Moreover, the level of the needs assessment will be determined by available financial and other logistical resources for potential program implementation. The process of data collection requires intensive training on proper management of the needs assessment, resource concentration, sufficiency in resources availability, and accuracy and commitment. The analysis of data collected by needs assessment in qualitative and quantitative methods should be based on categorizing the level of high risk behaviour. Optimal program planning, monitoring and evaluation will follow.

In discussing communication and mine risk education, the focus was on interpersonal, small and traditional and mass media, community liaison channels, as well as mine risk education strategy for the community. In order for any mine risk education program to be effective it should use different kinds of communication processes such as reading, writing, dialogues, questions and answers, TV, and learning in the classroom. In this respect, communication techniques such as using the voice, facial expressions, and movement, and communication media play a major role. These can be categorized into different types including interpersonal, small media, traditional media, mass media, community liaison, and education communication strategies. It is essential to have a communication strategy which is based on communication processes, techniques and channels that are most suited to specific audiences.

Notes

[1] See http://www.gichd.org/main-menu/about-gichd/annual-reports, Accessed on Aug, 2, 2011.

CHAPTER FOURTEEN

LEGAL ACTIONS AGAINST LANDMINE USERS

War and defence machines were considered to be one of the tools associated with the evolution of the international community and the emergence of the industrial revolution. It was also associated with the ambitions of colonial wealth in the world in addition to the dream of controlling people and their resources and the coveted power states in controlling the wealth of other nations. The power states have used modern war and destructive tools to achieve their goals, and from World War II started an arms race and developed new instruments for death and destruction.

Conventional weapons were considered the most tragic sources of the consequences of wars through the history of mankind. According to Themnér and Wallensteen (2011), 266 armed conflicts occurred during the period 1989 to 2010 and most of the victims were civilians. At the beginning of the twentieth century, 10% of the victims of armed conflicts were civilians and this ratio was raised since 1989 to constitute as much as 90% of the victims. All these victims resulted from the use of conventional weapons that are excessively injurious and have indiscriminate effects.

The expansion of the use of mines and placing them randomly has led to a huge humanitarian disaster targeting both humans and livestock. It also poses a serious threat to the environment and causes the inability to achieve integrated and sustainable growth.

The problem of landmines and especially the anti-personnel mines reached an extent considered a crisis on the global level. According to the landmine monitor reports in 2009 and 2010, there are more than 110 million mines laid in about 73 countries in the world, implying that on average there is one mine per one dozen children. Out of 77,290 landmine casualties since 1999, 18,908 were killed, 54,566 injured and 4,058 are of unknown status.[1]

A regional breakdown of the total global casualties recorded by the landmine monitor report from 1999 to 2009 is set out in the Table 14.1 below.

Table 14.1. Casualties by Region 1999–2009

Region	No of States	No of States with Casualties	No of Causalities
Asia-Pacific	40	21	35,780
Africa	48	32	16,924
Middle East and North Africa	18	17	8,882
Americas	35	14	7,890
Commonwealth of Independent States	12	12	4,638
Europe	42	23	3,176
Total	**195**	**119**	**77,290**

Source: Combination of the figures of landmine monitor reports of 2008 and 2009

Despite the enormous resources used in mine clearance, the prevalence of landmines is a major challenge facing the world. It is worth mentioning that on average there is a landmine accident every twenty-two minutes, causing seventy causalities per day. With the current rate of mine clearance and under the conditions of no additional placement of mines, it would take 1,100 years to free the entire world of mines.[2]

The period following the Cold War and the emergence of uni-polarity witnessed a significant international movement targeting most types of conventional weapon violations on the global level. This included items such as light weapons, anti-personnel mines and cluster munitions that led to the amendment of Protocol II of the 1980 convention relating to the use of landmines, booby traps and other devices.[3] This international movement led to a first comprehensive agreement to end the presence of anti-personnel mines from the arsenal of any nation as part of the agreement called the Ottawa Convention in 1997. It is the first disarmament convention in history that called for a weapon to be banned in a comprehensive manner, and was followed by the Dublin Convention that banned cluster munitions in 2008.

The international humanitarian law has played a key role in adding a number of principles that have dealt with the whole legality of the use of certain weapons (Carnahan 1984). These principles are essentially based on two ideas: first is that humanity should not be overlooked by the military and political decision makers, and second is the idea of security and stability of states, which is usually expressed as a military necessity.

Human history has witnessed different stages and periods in this regard. The custom international humanitarian law has not restricted the methods and means of warfare and without exception, as long as it is considered as a means of defence of national sovereignty. Here, the interest of supreme states that required the use of fighting have stemmed from the wills of the leaders (Bugnion 2002). This idea was represented especially in the nineteenth century on a significant doctrinal basis without considering the effects of excessive damage and unnecessary suffering that accompany the use of certain types of weapons. The atrocities committed against humanity during that period of time called for the urgent need to review those principles in accordance with the principles of civilised humanity that refuses to unleash warring forces.

The conflict between the two ideas described above has resulted in the formulation of another international principle, namely the concept of compromise between the ideas of humanity and military necessity. This is called the principle of proportionality in the use of methods and means of warfare which requires parties to use them in way that is compatible with military objectives without leading to excessive injurious effects on soldiers and indiscriminate effects on civilians (De Mulinen 1989).

From this point, we will discuss first the principle of military necessity in Section 2 and then the principle of proportionality in Section 3, and address the most important international instruments that dealt with the two principles in a regular manner. We attempt to apply them on the legality of using landmines and cluster munitions in Section 4, while Section 5 elaborates the Kurdistan region in relation to the international landmine law and its application.

1. The Principle of Military Necessity for the Use of Landmines and Cluster Munitions

The legality of the use of landmines and cluster munitions has been subject to many jurisprudential debates. A simple review of the armed conflicts that have occurred over the past centuries will undoubtedly show that the warring parties justified the use of certain types of weapon based on the existence of a fair cause that led them to use it without care of the accompanying atrocities against the humanity. These acts have often been based on the belief of denial of any mercy for enemies since they represent injustice and unfair causes, and hence are responsible for the massive destruction and suffering of the people on the both sides.

The concept of military necessity has evolved along with the concept of justice known from the early times of human history, and was a legitimate basis for all wars. The international humanitarian law has contributed in codifying many of the prevailed customs through history to curb the actions of those leaders who have justified the use of weapons without discriminating between the military and civilian population (Grant

2003). The positive contribution is the far from effective elimination of such threat or the significant reduction of its destructive impacts.

The just war theory, as a method of evaluating military actions, has been recognized historically by thinkers such as St. Augustine of Hippo[4] and St. Thomas Aquinas.[5] They believed that military necessity is only a part of insurgency and went so far as to say that there is no separation between them, and combat operations are the will of heaven delegated to the king to execute; the just war therefore stems from the existence of necessity. Among the conditions required for a war to be just, as St. Thomas Aquinas stated, are: power legitimacy and will, existence of a just cause, and achieving positive goals and avoiding devilish deeds (Grant 2003).

However, other thinkers added other conditions to those listed above, such as the imminent risk of threats to the existence of a nation, and the combat operations sufficiently necessary for the nation's interest. Under such conditions, the combat operations in self-defence should be based on methods and means that are suited to the threat that faces the nation.

From the legal aspect, the concept of military necessity has been addressed in different international levels and on the national level. For instance, some countries have committed their armed forces to the necessity of taking appropriate action prior to taking any military action. The British army field manual defines these appropriate actions as military necessity authorizing the belligerent party who use methods and means of warfare that are not banned by the law of armed conflict. The use of such methods leads to the achievement of the legitimate purposes with an emphasis on completing the requested task either partially or completely against the enemy as early as possible with little loss of life and property (Meyrowitz 1994).

In terms of international humanitarian law, the concept of military necessity has been mentioned in many international instruments. It was first mentioned in the preamble to the declaration of St. Petersburg in 1868. Another instrument is the drafting of the declaration by the participating countries that have been convinced to include this important topic on the conference's agenda. The inclusion stating that the only legitimate object which must be sought by the countries during wars is to weaken the military forces of the enemy, and the Hague Convention of 1907 confirmed this principle in its preamble (Meyrowitz 1994).

Based on the above discussions, it can be concluded that when countries use the principle of military necessity, they aim to evade the international responsibility of the violation of international humanitarian law. In order to curb this abuse, or at least to mitigate it, international humanitarian law has recently agreed on considering military necessity as the only acceptable response to military operation, and some international treaties have a specific condition for using it such as Geneva Convention (IV) relating to the Protection of Civilian Persons in a Time of War.[6]

According to the customary international humanitarian law and a number of international treaties, there are some conditions which justify military necessity as follows:

- To achieve legitimate military objectives which cannot be done by only usual military procedures.
- The results from such military procedures on civilian losses to be as low as possible (Grant 2003). This is the protocol additional to the Geneva Convention of 1949 relating to the protection of victims of international armed conflicts (protocol I) of 1977, stated in chapter III, civilian objects. Here follows article 52, on the general protections of civilian objects, paragraph 2:

> "Attacks shall be limited strictly to military objectives. In so far as objects are concerned, military objectives are limited to those objects which by their nature, location, purpose or use make an effective contribution to military action. In this context it refers to those whose total or partial destruction, capture or neutralization, in the circumstances ruling at the time, offers a definite military advantage."[7]

The countries that have supported the use of anti-personnel mines have had a conservative position on the legality of the use of this weapon, especially if we know that the international principles in international humanitarian law are incompatible with the effects of landmines and cluster munitions. One of these principles is the annex to the Hague convention of 1907: Regulations respecting the laws and customs of war on land, Section II: Hostilities, Chapter I: Means of injuring the enemy, sieges, and bombardments, article 23: In addition to the prohibitions provided by special Conventions, it is especially forbidden, to destroy or seize the enemy's property, unless such destruction or seizure be imperatively demanded by the necessities of war.[8]

In spite of the fact that the supporters to the use of landmines rely on this principle, this argument is not be strong for two reasons. First, landmines, with all their specifications, are a weapon which cannot be accurately directed against the enemy. Second, the continuous threat for long periods of time classifies this weapon as beyond the concept of military necessity (Sandoz 1981).

Further evidence of the illegibility of using landmines as a military necessity is stipulated in paragraph 3 of article 2 of the Protocol on

Prohibitions or Restrictions on the Use of Mines, Booby-Traps and Other Devices, as amended on May 3, 1996 (Protocol II to the 1980 Convention as amended on 3 May 1996). This convention states that: "It is prohibited in all circumstances to use any mine, booby-trap or other device which is designed or is of a nature to cause superfluous injury or unnecessary suffering."[9]

Moreover, because landmines are characterized by these specifications, the ban and lack of legitimacy is stronger than the evidence for their permissibility and legality. Furthermore, the phrase "prohibited in all circumstances" extends its prohibition even in the context of military necessity.

Through applying the conditions of military necessity to the use of landmines, particularly the condition of achieving legitimate military objectives, it can be said that this weapon in its devastating affects exceeds definite military advantage. This is supported by Paragraph 2 of Article 52 of Protocol Additional to the Geneva Conventions of August 12, 1949, and relating to the Protection of Victims of International Armed Conflicts (Protocol I), June 8, 1977 that states that:

> "Attacks shall be limited strictly to military objectives. In so far as objects are concerned, military objectives are limited to those objects which by their nature, location, purpose or use make an effective contribution to military action and whose total or partial destruction, capture or neutralization, in the circumstances ruling at the time, offers a definite military advantage."[10]

Moreover, it is illogical to accept landmines as a weapon for military necessity due to the lack of proof for the capacity and the effectiveness of this arm in military operations rather than for defence plans, and the evolution of military industry has allowed for the development of sophisticated weapons that can be easily replaced at a lower cost than landmines (Davis and Duniop 1994).

The after effects of using landmines in armed conflict areas show a concept and a vision different from that of military necessity. The outcome will call at least for the ban or restriction on the use of this weapon according to the principle of humanity as it represents a philosophy of international humanitarian law. The law was created and established as a call for establishing rules for the prevention and not for the legalization of landmine use labelled as military necessity.

2. Principle of Proportionality

We will discuss in this section two related subjects. First, the reasons for the emergence of the principle of proportionality, and secondly a study of the most important explanations of the principle and its relationship with the legality of the use of landmines and cluster munitions.

2.1 The Evolution of the Principle of Proportionality

The emergence of the principle of proportionality was not a coincidence, but the inhumane results of the methods and weapons used in armed conflicts both internationally and internally. The main reason is the principle of proportionality is the eternal conflict between the two opinions. The first calls for the use of all methods or means of combat that achieve the military objective, the roots of which go back to the theory of just war mentioned in previous sections. The second calls for the avoidance of incidents involving those who have nothing to do with military operations, namely civilians, as well as calls to reduce the pain and injuries between fighters. This principle has its roots in the principles of religions like Judaism, Christianity and Islam.

The core of the abovementioned principle is valid when applied in general to armed conflicts with the objective of mitigating the horrors of war by emphasizing the human side during armed conflicts. It should be mentioned that the equation underlying the principle of proportionality is very complex as it is based on the approximation and harmonization between the good performance of military operations and the achievement of military objectives on the one hand, and respect for humanitarian principles referred to by international humanitarian law on the other (Engle 2009).

2.2 Explanations of the Principle of Proportionality

The history of international law witnessed the first international legal document represented in the language of a notable declaration in this regard (the St. Petersburg Declaration of 1868). The international humanitarian law is designed to "conciliate the necessities of war with the laws of humanity."[11] As Beach (2001) pointed out, the St. Petersburg Declaration has emphasized the need to review the legality of the use of any weapon about which doubts might be raised regarding compatibility between the necessities of war and humanitarian principles.

The different theories of military doctrines have caused variations in permissibility of using methods and means of combat, and hence the role of international humanitarian law is important in which it seeks to reduce losses and avoid humanitarian suffering as much as possible. In this regard, a number of conditions, if they are available, will justify military operations based on the principle of proportionality, and are stated as follows:

- A complete control of the military leadership's decisions and the sources of fire to prevent any serious violations of international humanitarian law.
- Sufficiency of necessary military operations to defeat the enemy—for example, to destroy 60% of the human and military capacity of the enemy is enough to overcome and defeat it, therefore there is no need to destroy the rest of the personnel and equipment.
- Disallow ordering or pre-planning an attack that may lead to human genocide.
- Do not make indiscriminate attacks such as those that are not directed at a specific military objective.
- Refrain from military operations that cause unjustified pain or injury.
- Do not make attacks against the civilian population or civilian objects.
- Strive to direct all operations and the sources of fire on military targets rather than civilians, whether accidental or indirectly.[12]

For landmines, the principle of customary international humanitarian law is the first to organize its uses. Several problems have been raised about incorporating the effects of this weapon with the military advantage and the effects of inhumane consequences raising question marks about the legality of their use. Considering that this weapon cannot be limited to short-term effects on civilians and also on combatants, to curb those effects the proportionality of use has been organized for the first time through the Protocol on Prohibitions or Restrictions on the Use of Mines, Booby-Traps and Other Devices as amended on May 3, 1996 (Protocol II to the 1980 Convention as amended on 3 May 1996). The protocol prohibited the use of landmines in certain circumstances, and in article 3 the general restrictions on the use, of mines, booby-trap and other devices are outlined, and the description of indiscriminate use is discussed in section 8:

> "The indiscriminate use of weapons to which this Article applies is prohibited. Indiscriminate use is any placement of such weapons:
>
> (a) Which it is not on or directed against a military objective. In case of doubt as to whether an object which is normally dedicated to civilian purposes, such as a place of worship, a house or other dwelling or a school, is being used to make an effective contribution to military action, it shall be presumed not to be so used; or,
>
> (b) Which employs a method or means of delivery which cannot be directed at a specific military objective; or,
>
> (c) Which may be expected to cause incidental loss of civilian life, injury to civilians, damage to civilian objects, or a combination thereof, which would be excessive in relation to the concrete and direct military advantage anticipated."

Moreover, a few questions can be raised: can landmines be directed at a specific target? Is it possible to make precautions to avoid this weapon when it is out of control and inflicting civilian causalities? If it is expected for this weapon to be out of control, is there any proportionality between the injury and the concrete and direct military feature?

Studies and researches on landmines confirm that this weapon cannot be directed at specific military objectives except in rare cases, and the inhumane effects on civilians and civilian objects surpassing the military advantage that is usually justified by military commanders during the armed conflict. Based on these arguments and facts, thereafter, the following question can be raised: Is it possible to restrict the use of landmines to the extent appropriate and consistent with the requirements of humanity, and approved by the customary rules of international humanitarian law?

Addressing the disproportionality that characterizes the relationship between landmines and the principle of proportionality has put forward several perspectives. These are towards believing that the expected inhumane effects of landmines can be reduced, calling for partial treatment of these effects with continuous support of using them. These perspectives are divided into three groups. First, to ban certain types of landmines, such as non-detectible mines which explode when detected. Second, calls for mitigating the impacts of mines through adding special techniques that allow for self-neutralization mechanisms or self-destruction. Third, calls for the need to review the UN agreement on Conventional Weapons of 1980, known as the Convention on Prohibitions or Restrictions on the Use of Conventional Weapons, which may be deemed to be excessively injurious or to have indiscriminate effects, by adding new provisions that restrict their use.

The Protocol on Prohibitions or Restrictions on the Use of Mines, Booby-Traps and Other Devices as amended on May 3, 1996 (Protocol II to the 1980 Convention as amended on May 3, 1996), Article 3: "General restrictions on the use, of mines, booby-traps and other devices," has committed the states who use landmines to take all the necessary and feasible precautions to protect civilians from their effects. It defines the possible precautions as:

> "10. All feasible precautions shall be taken to protect civilians from the effects of weapons to which this Article applies. Feasible precautions are those precautions which are practicable or practically possible taking into account all circumstances ruling at the time. These include humanitarian and military considerations. These circumstances include, but are not limited to the following:
>
> (a) The short- and long-term effect of mines upon the local civilian population for the duration of the minefield.
>
> (b) Possible measures to protect civilians (for example, fencing, signs, warning and monitoring);
>
> (c) The availability and feasibility of using alternatives; and
>
> (d) The short-and long-term military requirements for a minefield."

Despite the international efforts made to mitigate the expected impacts of the use of landmines, and especially those related to the injuries to civilians and civilian objects, these efforts and proposals have failed to achieve the stated goals. Such mechanisms or restrictions on use have not been observed, especially in non-international conflicts. Thus, there will not be any consideration or precautions that may be taken to reduce its inhumane effects (Pictet 1966). However, based on the above mentioned we believe that the principle of proportionality does not fit at all with the use of landmines.

3. Words of Caution about Mine Action Responsibilities

The mine action activity in Kurdistan was previously financed and managed by the United Nations administered Oil for Food Program and some other independent Aid agencies. The Iraqi National Mine Action Authority was established in 2003. It has taken responsibility for strategic planning, budgeting, coordination of different projects and activities, coordination of relations with donors, and setting national mine action standards. Among other tasks are maintaining the national information management system database for mine action. This has exposed Kurds to a higher risk level and subjected them to an Iraqi legal system which is not in their favour.

The transfer of financial, manpower and action responsibilities from international NGO and regional authorities to the central authorities is unfortunate. It has placed the Kurds in a vulnerable situation. In addition, the reorganization has resulted in inefficiency, low intensive or lack of clearance activities, and the continued suffering of the civilian population in the region. The deteriorated situation is a direct result of the inability of the Kurdistan regional government to negotiate a better solution and to allocate necessary resources financed by regional budget. Furthermore, formal legal claims should be made such that the cost of mine clearance and victim rehabilitation be covered by the central government budget.

Together with the start-up of mine action activity in the Kurdistan Region in 1999, UNOPS established a comprehensive information management system. The system allows for the management of all mine related data processing. All previous minefield information records since 1992 was imported to the new system. The new information database was originally developed and designed to describe the overall mine and ERW problem in the region. It needed adaptation to further developments in the mine action sector and helped to implement and integrate information aspects to adhere to the new international mine action standards for humanitarian mine clearance operations. Certain improvements such as verification of the quality of the existing data, management of missing fields and modification of the structure of the database was also undertaken. In sum, the process described above did allow the Kurdistan region mine action program to enhance its strategic management and its priority setting capability.

The new system involves reporting data from regional to central mine action authorities. The same database is used by both parties for planning and managing mine action operations. All information entered into this new system is automatically reflected in the Baghdad database and should be available to plan future operations. However, care should be taken such that this sensitive and detailed information is not available to the Iraqi military and used in future mine activities.

The regional government is urged to allocate a fixed share of its annual budget which is raised from oil and gas revenue sharing to landmine awareness and clearance, and landmine related health care provision and rehabilitation programs. This will make the region independent of foreign NGOs with different priorities. Planning and management of mine action in the region should be strictly in the hands of local people. Currently the stalled mine action activity is in the hands of the central government authorities, part of which in the past authorized minelaying in the region.

Thus, a centrally nominated and financed authority is not trustworthy to lead the clearing actions in the Kurdistan region.

Although Iraq has recently signed the international anti-personnel mine treaty, none of the minefield records have been made available to the Kurdistan region's mine action program. The lack of minefield records has made the minefield clearance costly and lengthy. It indicates an unchanged central government policy towards the Kurdish population and its regional government. The regional landmine agency should be cautious in distribution the collected data to the central government landmine agency which provides detailed information about locations of landmines and their impacts. The information is too sensitive to be accessed by the army and security forces of Iraq and there is no doubt that these forces will use the information in future mine-laying activities in the region.

4. Policy Recommendations

We suggest a number of recommendations on enabling the Kurdistan regional government to sue the Iraqi government as an illegal user of landmines and to sue the countries who supplied the landmines. Due to the complicated nature of the issue, KRG is required to establish a legal committee to raise the issue of Kurdish mine victims in a way that secure their families financially and force the Iraqi government to allocate a necessary budget for those victims. Furthermore, the KRG should make a serious effort to implement a landmine law and force the Iraqi government to legislate it, such that the victims and internally displaced people are protected and compensated for their sufferings.

Alongside of the legal committee, we suggest a research institute be established with the primary objective of conducting research on landmines to analyze collected statistics. Such institutes will play a major role in coordination of research related to different aspects of landmines such as information, prevention, training, disability, rehabilitation and the labour market. In addition, it will enable the gathering of evidence on the crimes committed which support the regional government in following up the landmine issue for matters of prevention, reducing impacts, victim compensation and bringing the landmine users to justice.

After agreement on a united but federal Iraq, the landmine responsibility was transferred from international organizations and local agencies to the central government. From the Kurds' perspective the transfer of responsibility to a central mine action agency to plan and implement demining is a major step backwards. This is evidence of a lack of awareness and interest of the Kurdistan regional government to the issue.

This action seems to be inappropriate considering the Iraqi government's crimes against Kurds, including the systematic laying of mines in the Kurdistan region. There is likelihood that the collected information might be misused by the Iraqi army and security forces in a future conflict. The regional government should promote local capacity and through its legal system, or an international legal system with reference to international law related to landmine use, persecute those responsible for laying mines in the region. The KRG should do their utmost to force the central government to allocate a special budget to clean minefields and to rehabilitate the victims of their crime.

5. Conclusions about Legal Actions against Landmine Users

The inequality in the use (and overuse) of some types of conventional weapons has been the main features of armed conflicts in the current era. Although it is not a new phenomenon at all, it is important to consider the related principles of international humanitarian law in the use of certain conventional weapons in the absence of internationally clear expressions on this issue. It is no longer sufficient to refer only to these principles in the core of international instruments without finding a mechanism to curb the violations that characterize contemporary armed conflicts, especially after the 2003 war against Iraq and the war in Lebanon in 2006 that were characterized by serious and frequent violations of the principles discussed above. The principles refer only to use of landmines, but not to their production and supply.

In this chapter we have addressed two important principles of international humanitarian law related to the use of certain conventional weapons, their military necessity and the proportionality and adaptation of the use of certain types of conventional weapons. In this regard, landmines and cluster munitions are to be considered the most hideous and barbarous weapons that have known to the military arsenals today. It is clear from what has been put forward in this chapter that in spite of regulation by the international humanitarian law regarding the major issues related to armed conflict, it has not been effective or sufficient in dealing with the terrible events associated with these armed conflicts.

Countries and non-international parties in armed conflicts often try to justify serious violations based on the principle of military necessity and proportionality. Because there is no international agreement to identify a mechanism in which it determines the legitimacy of using certain types of conventional weapons, the conflict parties will continue to violate the international humanitarian law. This is prevalent in cases where catastrophes and atrocities continue to kill many civilian people who have no direct relationship to these conflicts.

In order to curb these serious violations of international humanitarian law, there must be a specific mechanism to investigate these violations and to prosecute those responsible. This can be done through establishing a special international court which deals with violations of international humanitarian law to act as a deterrent to violations that the developing countries face amid the silence of the world's great powers and producers and suppliers of such inhuman weapons.

The countries responsible for the formation of such a court should be the same countries who led the negotiation in Ottawa and Oslo as they contributed to two important humanitarian agreements, namely the Ottawa Convention to ban landmines in 1997, and Dublin Convention of 2008 to ban cluster munitions. In the process, they should take advantage of the experience and evidence of atrocities in the countries suffering from the use of such inhuman weapons. The major powers with access to technical, financial and organizational capacity to contribute to the formation and operation of such a noble activity but with veto power in international institutions seem not be legitimate partners. They have shown, in practice with no exception to take sides with their businesses. Under such conditions and their strong influence, it seems that many improvements along the lines discussed above, which favour the suffering civilian populations, are not forthcoming.

Notes

[1] Landmine Monitor Fact Sheet (2010), "Impact of Mines/ERW on Children." Available at: http://www.the-monitor.org/index.php/LM/Our-Research-Products/Factsheets, Accessed on Aug, 12, 2011.

[2] See chttp://www.uncg.edu/student.groups/sal/Pages/facts.html, Accessed on March 1, 2011.

[3] See http://www.icrc.org/ihl.nsf/FULL/575?OpenDocument, Accessed on March 1, 2011.

[4] See http://www.newadvent.org/cathen/02089a.htm, Accessed on March 1. 2011.

[5] See http://faculty.cua.edu/pennington/Law111/AquinasJustWar.htm, Accessed on March 3, 2011.

6 Article 53 stated that "Any destruction by the Occupying Power of real or personal property belonging individually or collectively to private persons, or to the State, or to other public authorities, or to social or cooperative organizations, is prohibited, except where such destruction is rendered absolutely necessary by

military operations", International Humanitarian Law—Treaties & Documents, ICRC.
[7] The full text of preamble is available online at:
http://www.icrc.org/ihl.nsf/full/470?opendocument
[8] The full text of preamble is available online at:
http://www.icrc.org/ihl.nsf/WebART/195-200033?OpenDocument, Accessed on July 1, 2011.
[9] The full text of preamble is available online at:
http://www.icrc.org/ihl.nsf/WebART/575-03?OpenDocument, Accessed on July 1, 2011.
[10] The full text is available online at:
http://www.icrc.org/ihl.nsf/full/470?opendocument, Accessed on June, 27, 2011.
[11] The Full Text of the Declaration preamble is available online at:
http://www.icrc.org/ihl.nsf/INTRO/130?OpenDocument, Accessed on July 1, 2011.
[12] The International Law related to military operations, available at:
http://www.icrc.org/ihl, Accessed on Aug. 12, 2009.

CHAPTER FIFTEEN

SUMMARY, CONCLUSION AND POLICY RECOMMENDATIONS

1. The Origin of Landmines

Mines are explosive monitions that are placed under, on, or near the ground to be triggered by the presence, proximity, or contact of a person or a vehicle. There are two basic types of mines: anti vehicle or anti-tank mines and anti-personnel mines. The history of landmines can be traced back to the American Civil War in the nineteenth century where they were referred to "Land Torpedos." Mines as they are known today were initially developed during World War I following the introduction of tanks in warfare. Large anti-tank mines were developed to counter this new invention and the much smaller anti-personnel mine was sown throughout anti-tank minefields to deter enemy soldiers from entering. Landmines became a prevalent weapon on the battlefield during World War II. Landmines are victim-activated and indiscriminate in their targets and impact. One of the most effective anti-personnel landmines during that time was the German-made "bouncing betty," which was designed to jump from the ground to hip-height when activated and to propel hundreds of steel fragments within a wide range.

The main characteristic of mines unlike other weapons is their unique time-delay function, as they are not designed for immediate effect. Mines can stay inactive for an almost unlimited time until triggered by a victim. This unique and inexpensive device has been used intensively during wars between countries and in internal conflicts to block access to large piece of land. During the time between the two world wars, mines became easier to access and their design was modified. Landmines are technologically very simple devices to manufacture, and China and other nations are bale to manufacture mines at low costs. Mines are made mainly from metal, however, plastic, ceramic and wooden landmines are also designed to make detection difficult.

Landmine contamination worldwide can be considered a serious and hazardous issue to humankind. There is an estimation of 100 million anti-personnel mines in the ground and another 100 million stockpiled around the world. According to United Nations reports, if the use of mines were to be stopped immediately, it would take another one hundred years and a cost of more than $33 billion dollars to clear the existing landmines in use. The existence of mines is considered to be a serious socio-economic and environmental problem facing many countries exposed to their use. Landmines and unexploded ordnance are a significant threat and barrier to economic and social development in many parts of the world suffering from internal and external conflicts.

The prevention of access to land for agricultural use, transportation, and also blockages to water access have caused the displacement of people in many mine-affected countries. Mines also cause the absence or low development of basic infrastructure in those affected places. They also cause land and soil degradation and limit agriculture productivity. The solution to these problems and reduction of their impacts are the main purpose of the mine action activities. Landmines pose a great risk of death and injury to civilians and the land. Mines typically maim or kill the most productive members of a community, and prevent refugees and internally displaced persons from returning to their homes after conflicts have ceased. They cause severe poverty and underdevelopment in contaminated regions and block access to land and limit most socio-economic activities in rural areas. Beside the direct effects on life, landmines impose a heavy economic burden on these communities in the forms of medical, psychosocial, political and economic impacts.

Mine action is the term used to describe all activities geared towards addressing the problems faced by populations as a result of landmine contamination. Its aim is not limited to surveying, marking and eradicates landmines but also encompasses humanitarian and developmental activities. The objective is to recreate an environment in which people can live safely, and economic, social and health development can occur freely.

Mine action comprises five complementary groups of activities. These include: mine risk education; humanitarian demining surveys; mapping, marking and clearance; victim assistance including rehabilitation and reintegration; stockpile destruction, and advocacy against the use of anti-personnel mines. Thus, mine action involves activities which focus on prevention of the future use of landmines, reducing the risk of injury from existing mines and assisting injured individuals. Preventing the future use of landmines is carried out through international campaigns. Reducing the risk of injury to people living in contaminated communities is conducted through mine risk education and demining programs. Victim assistance is the provision of assistance for survivors of landmine incidents.

Mine action is normally just one of many activities taking place in a post-conflict environment and often a foundation for subsequent humanitarian and developmental initiatives. It is also an integral component of strategies designed to rehabilitate health care, education, infrastructure and agriculture required by societies to recover from violent conflicts. The two main types of demining activities are military and humanitarian. Military demining only clears the required pathway for the troop's movement. Humanitarian demining aims to clear land and thereby allow the civilians to return to their lands to perform their daily life activities. The aim of humanitarian demining is to restore peace and security at the community level.

Humanitarian mine action operations face many complications on the ground. There are various aspects of recovery to be addressed. These aspects include peace building, repatriation of refugees, rehabilitations of communities, reconstruction and development. For planning purposes, several stages can be identified between the extremes of open conflict and stable self-dependency. During open conflict, landmines may significantly constrain the movement of peacekeeping forces and humanitarian personnel. In addition to focussing on supporting peacekeeping and emergency humanitarian activities, mine action supports rebuilding indigenous organizations. A crucial period of transition exists between the humanitarian emergency and assisted development phases. A key characteristic of the assisted development stage is the existence of a central governing authority operating under conditions of security, democracy, and self-dependency.

There are various stakeholders in mine action, including those internal and those external. The power and interest of each will vary as the country progresses towards the achievement of stable self-dependency. Mine action is normally only one of many activities required in a post-conflict environment, but it is often the foundation for humanitarian, rehabilitation, and development initiatives. It must also be considered as an integral component of strategies designed to rehabilitate health care, education, infrastructure, agriculture and other systems. The mine action environment is quite complex and it brings together many stakeholders from different disciplines to cooperate in order to achieve same stated goals. Its success depends on the ability of managers to coordinate and effectively plan the different activities.

In the Kurdistan region, landmines will continue to be a major obstacle to the development and economic growth in rural border communities.

The objective of this work was to investigate the magnitude of the mine problem in the region, the current mine action activities as well as those of the future to clear the region to facilitate rehabilitation and development programs. In addition, this book emphasized the social, health, education and economic impacts of mines and their legal consequences.

2. The International Scope of the Mine Problem

The issue of landmines is becoming more and more sophisticated as technology development allows for further and more profitable experimentation. The demand for their use remains high as many countries are in conflict with neighbours as well as minorities within their territories with political, social, and cultural demands that are not satisfied by undemocratic and often non-elected governments. These governments use anti-personnel mines as effective and low cost measures in response to legitimate demands and as a means of conflict resolutions.

The international scope of the mine problem and actions undertaken to prevent their use or to reduce their impacts was discussed with the main objective of attention to the conditions under which innocent populations are suffering from their frequent, random and sometimes calculated uses. Despite growing international interest among governmental and non-governmental organizations to ban the use of mines, there is no interest in banning or limiting their production, in particular when anti-personnel mines is considered. The combination of technology development and the high frequency of domestic and cross-border conflicts have made mines an international problem requiring global action to eliminate this inhuman phenomenon. Here, these issues were discussed and some statistics presented to show the magnitude of the problem, its regional distribution and negative socio-economic effects.

There is a strong movement against mines, the progress of which is halted by major producers and users of landmines, namely the USA, Russia and China, all of which are members of the UN Security Council with right of veto. These countries do not hesitate to use their veto for their own national economic, militarily and strategic interests. This situation has in many cases shown to be a major obstacle in conflict resolution between countries and other issues of global character. Inefficiency and the bureaucratic nature of existing international organizations dominated by the major powers adds to the handicapped situation in bringing the necessary changes and the achievement of goals to enhance the well-being of humankind.

3. The Scope of the Mine Problem in the Kurdistan Region

Iraq in general and the Kurdistan region in particular is heavily contaminated with explosive remnants of war, landmines, and unexploded ordnances. The mines are related to the frequent wars with neighbouring countries, the 2003 US invasion and the Iraqi government's policy towards the Kurdish population. In the Kurdistan region, mines were laid during four distinct periods. The first was in the mid-1970s, when the central government attempted to subdue the Kurdish resistance. The second was during the Iraq-Iran war, when the border areas were mined. The third was during the Gulf War, when barrier minefields were laid along the border with Turkey. The fourth was in the aftermath of the Gulf War, when Kurdish factions used landmines in their internal conflict.

Although Iraq has recently signed the international anti-personnel mine treaty, none of the minefield records have been made available to the Kurdistan region's mine action authority. The lack of minefield records has made clearance costly and lengthy. It indicates an unchanged central government policy towards the Kurdish population and its regional government. The mine action activity in Kurdistan was previously financed and managed by the United Nations administered Oil for Food Program and some independent aid agencies. The Iraqi National Mine Action Authority was established in 2003. It has taken the responsibility for planning, budgeting, projects' coordination, donor's relations, setting standards and maintaining the national mine action database.

The transfer of financial, manpower, information management, and different action responsibilities from regional authorities to the central authorities is unfortunate and has placed Kurds in a vulnerable situation. In addition, the reorganization has resulted in inefficiency, low intensive or lack of clearance activities and the continued suffering of the civilian population in the region. The deteriorated situation is a direct result of the inability of the regional government to negotiate a better solution and allocation of necessary resources financed from the regional budget. Furthermore, so far no formal legal claims have been made to cover the cost of mine clearance and rehabilitations from the central government budget.

4. Methods of Dealing with Mines

The number of landmine victims is large and increasing due to the intensity in the use of landmines, technology development in making them undetectable, their stronger impacts, population increase and increased demand for food and energy. Landmines are designed to cause death or serious injury to humans. A landmines victim suffers mostly from the shattering effect as well as the subsequent medical and social burden. Through contamination of useful and fertile land areas, it is an effective obstacle to the development of the societies. Mines put civilians at risk of death or serious injury and create long-term harm and destruction in particular to rural areas with limited resources. The presence of mines in agricultural lands can cause famines and the increased need of food aid. It leads to the migration of the rural population and affects the concentration of the population in urban areas and subsequent increased public service supply shortages.

The impact of landmines on individuals and communities goes beyond physical effects to include psychological, social and economic effects. Children suffer most from the limitations on their free movement. The increasing awareness of the global impact of landmines on civilians was a result of the establishment of humanitarian mine clearance programs. Efforts are needed to better organize the humanitarian mine clearance at international level and to develop criterion, standards and procedures that are effectively adoptable to further enhance their effectiveness, reliability and safety in practice.

Here we shed lights on the methods dealing with mine action activities. Mine action involves those activities which prevent the future use of landmines, reduce the risk of injury from existing landmines, and assist those already injured. The degree and form of risk from landmines and the nature of the mine action response varies from one location to another. Different landmines clearance steps and clearance methodologies employed along the complicated clearance process were discussed. The process involved, among others, general survey, technical survey, marking, clearance, and completion survey steps.

5. The Landmine Impact Survey

The Landmine Impact Surveys grew out of the collaborative efforts of the international campaign to ban landmines and the Ottawa treaty. They are aimed at meeting the overlapping needs of international donors, national authorities and humanitarian mine action planners and implementers. The overall vision is to facilitate the prioritization of human, material and financial resources supporting humanitarian mine action at the national, regional, and global levels. To fulfil this vision, the survey projects are executed to meet the same global standard. Their implementation facilitates an improved collaboration between the three primary stakeholders of the process: national authorities, donors and implementing agencies. Through the survey, useful data is collected and support is provided for further steps in the mine action process including marking, clearance and completion survey. In addition, it provides datasets for other purposes such as research.

The data collected during the survey process helps to define the entire problem, improves national planning efforts, fosters development of national plans, and establishes baseline data from which outcomes and performance can be measured. The surveys are considered to be the first and most vital step in the overall humanitarian mine action and intended to improve the quality of information available for decision makers at all levels. The survey data allows donors to rationalize their allocation of funds to the areas of greatest human needs, permits national authorities to develop national plans focusing upon regions and areas of greatest impact, and gives implementers baseline impact data that will provide successful indicators for mine action programs outcome evaluations.

As a global initiative to standardize available information regarding the landmine problems around the globe, the landmines impact survey process represents a successful effort to ensure conformity of methods, procedures and processes. To ensure quality of results, internal and external quality assurance mechanisms are imbedded in the survey process. Surveys are based on the best practices in the field and are executed to measure impacts upon affected communities in a uniform manner. Due to the lack of capabilities of clearing at international institutions, the affected countries are expected to allocate internal resources to prioritize the local activities of mine clearance. The international and NGO sourced resources are limited. The limited available resources could be strengthened by taxing the mine producer countries.

6. The Landmines Impact Survey in the Kurdistan Region

An introduction and overview of the mine action in the Kurdistan Region is provided in this book. The focus is on the original approach and how it evolved to form an integrated mine action approach and the resulting differences in their outcomes. A detailed summary of statistics covering the mine action resources, local capacity building, classification of fields and communities, and victims categorized by their gender, age and professions are provided. The mine action in the Kurdistan region is elaborated along with all the problems related to mine contamination. It focuses on the use of mines, mine awareness education programs, mine victims assistance, survey of mine suspected areas and mine clearance activities.

The basic task of the mine action centre is to plan and implement mine clearance operations in the region. To fulfil the tasks, numerous supporting tasks and human skill development and organization structure activities are performed. The program consists of three core components: clearance, mine risk education, and victim assistance. These core activities previously planned their work largely in isolation from each other. The aim of the integrated program in 2003 was to coordinate all activities in order to address the overall impacts on individual communities. In order to implement the integrated approach, teams were made up of different specialists who worked together with villagers to address the reduction of the landmine impact. The use of mine action information systems supported the implementation plan through collection, management and supply of vital data. Introduction and implementation of the integrated approach has changed work procedure and outcome and learning by doing.

The landmine impact survey (LIS) in the Kurdistan region was conducted by UNOPS-MAP in late 2002. It was then followed by a LIS for Iraq conducted over a two-year period. LIS confirmed and complemented the original general survey conducted within the region during 1998 to 2000 building on data collected during the previous years. The two surveys' comprehensive coverage served as complementary to each other and provided a detailed picture of the magnitude of the mine problem in the region. They were conducted to identify specific blockages to community social and economic activities.

Twenty-five to twenty-seven districts within the Kurdistan region are mine affected, within which are 97 to 124 sub-districts, 1,126 of 4,281 communities with an estimated total of 748,651 persons living in the impacted communities. The governorate of Suleimaniya contains the majority of mine-affected communities. In the design of the clustering system, focus was concentrated on a group of mine-affected villages. Village clusters were developed based on the classified mine affected villages as high, medium, and low. The lands were classified as non-agricultural uses, pasture, and agricultural areas. Pasture areas are the most affected land category by landmines, particularly around strategic hillsides and mountain tops serving as military strategic positions.

Completion of the survey resulted in a comprehensive set of mine related socio-economic impact data. This data allowed the mine action community to develop effective plans that target areas posing the greatest threat and impact. The survey results indicated that all three governorates of the region suffer from many adverse consequences of landmines contamination. The clusters of mine-affected communities correspond largely to the history of armed conflicts in which landmines were used. Mines were planted in the 1980s and 1990s and illustrate a random distribution of mined areas throughout the region.

The large scale of contamination in the proximity of densely populated areas has resulted in many accidents. The distribution of recent victims by age and sex indicates that 94.6% of the victims were males of which 44.8% were aged fifteen to twenty-nine. The most frequent categories classified by occupation involving mine accidents were children playing, shepherds, military, farmers and workers, accounting for 87% of all victims. Tampering with munitions is the most prominent cause of accidents. Regression analysis on landmine victims and causal factors reveals that the distance to the nearest community with recent victims is a better predictor of victim numbers than the distance to the border. The different blocked areas are further disaggregated into their sub-areas. Pasture, crop land, other uses of land and water usage areas dominate. Access to grazing land shows the highest frequency of blockage type at 98%.

7. Factors Influencing Mine Action in Kurdistan Region

In the Kurdistan region every year, many civilians are either killed or severely injured following contact with mines and ERW. The presence of mines and ERW has drained the scarce medical resources and hindered rural economic activities and production of food. The impact is strongest on the most vulnerable rural populations, and has hampered reconstruction and development efforts to bring peace and normality to their communities. The humanitarian mine action program in the region has faced difficulties in operating effectively due to a lack of funding, skilled demining personnel and the availability of technology to perform the demining in an effective way.

A transfer of funding, planning and implementation of mine action activities from the joint regional and international organization to the central government is a blow to the whole program as the latter was responsible for laying mines in the first place and as such not credible to fund, plan, implement and confirm demining activities. It will be hard for the suffering rural population to entrust the central agencies in taking responsibility for such a vital factor. The information is too sensitive to be accessed by the army and security forces of Iraq. There is no doubt that these forces will use the information in their future mine laying activities against the region. More appropriate would be that the central government takes responsibility by financing a third party to undertake the clearing and to provide rehabilitation and training programs and compensation to the victims and their families.

We have discussed the natural barriers to the demining operation in the Kurdistan region and the technical issues that influence the mine action process, its progress and outcomes. The practical consequences from the influential factors must be taken into account in appropriate stages. Each of these factors and their possible impacts on the mine clearance activities is briefly described. Education seems to play an important role in the context of mines and reduction of its threat and impact. It is desired that mine awareness programs take into consideration social factors such as poverty, gender and age which are influential. Measures are required to be effective and correspond to the heterogeneous characteristics and needs of different population groups such as refugees, internally displaced people and nomadic groups.

8. Management of Mine Action Information in the Kurdistan Region

Following the start-up of mine action activity in the Kurdistan region in 1999, UNOPS established a comprehensive information management system. The system allows for the management of all mine related data processing. All minefield information records from the mine action group database is imported to the new system. The new system was originally developed and designed to describe the overall mine and ERW problem in the region. The databases met the initial requirements but needed adaptation to further developments in the mine action sector. In order to upgrade the database to keep up with new trends and requirements, an assessment was made to determine which additional functionality needed to adhere to the new international mine action standards for humanitarian mine clearance operations. The assessment led to recognition of the need to make a number of adjustments.

The new system involved the reporting of data from regional to central mine action authorities and has enabled both parties to use the same database for planning and managing mine action operations. The additional data available provide better information to demining organizations. All information entered into this new system is automatically reflected in the Baghdad database each time database synchronization takes place between Erbil and Baghdad. The drawback of the new system is that sensitive and detailed information available to the Iraqi military is likely to be used in their future mine laying activities with an expected impact on the Kurdish civilian population. In sum, the process to restructure the database, verify, correct and collect additional data on the impact of landmines allowed the Kurdistan region mine action program to enhance its strategic management and its priority setting capability.

9. The Mine Action Survey Methodology

The mine action survey, program and composition of the mine action groups and their work procedure and data collection and management is described. The impact survey employed systematic data collection and analysis in cooperation with community representatives. The impact interview, mapping and assessment were conducted in all mine-affected communities. The results of the interviews were entered into a database, which formed the basis for subsequent analysis. Lessons learned during the survey resulted in additional training for the survey teams. The training was to ensure data consistency throughout the survey process. Management skills of the team leaders and location supervisors were also improved.

Based on earlier work conducted by the general survey teams, information was made available from the extensive mine action database covering all communities known to contain mines. A total of 2,133 key informants provided input to the data collection process. A survey team typically interviewed, surveyed and mapped a minefield in two days. Much of the time was spent interviewing the key informants and visual verification of individual mined areas. The group interview consisted of two parts: a confirmation of the mined site and a structured questionnaire interview. The questionnaire form contained three components including village form, minefield form and incident form. Finally, the data was entered into a centralized database and verified.

The result of the survey is overviewed in disaggregated form. The region is divided into its three governorates and further divided into different districts and by the magnitude of impact or frequency of accidents. The accidents are classified by the outcome and characteristics of the victims and the types of activities at the time of accident. The result indicates that the three governorates all suffer from adverse consequences of landmines and ERW contamination. The frequent existence of dense clusters of high and moderate risk villages suggests a well-planned and targeted set of mine action initiatives could make considerable improvements in the landmine situation in the region.

10. Analysis of Landmine Fatalities in the Kurdistan Region

The yearly cross sectional landmines victims' data from the early 1960s until 2003 in the Kurdistan Region was analyzed. In addition to descriptive analysis, we used regression analysis to identify determinants of the probability of getting killed by landmines, and to estimate the determinants' impacts. The individuals were observed when they are subjected to landmine accidents. In order to account for the dynamic aspect of the process and heterogeneity by location, a pseudo panel data was created where each of the districts was observed for the entire period. This allows for analysis of both heterogeneity and dynamics across locations and over time.

The empirical results show that males are more often subjected to landmine accidents than females, attributed to the nature of their activities. There is a negative age effect suggesting that the younger are more exposed to the risk of accidents. The most exposed groups are shepherds, farmers and Peshmarga soldiers. Walking is the type of activity that led to a higher rate of accident compared with touching and other daily activities. Awareness of landmine danger reduced the probability of accident but no significant reduction is observed from participation in training programs. A major part of the decline in the rate of accident is attributed to urbanization and migration from agriculture to service and industry sectors.

The population is still highly exposed to disproportionate danger of landmines harming the economy and social life, and causing enormous psychological suffering and pain. The regional government is urged to allocate a fixed share of its annual budget raised from oil and gas revenue sharing to landmine awareness and clearance, health care and rehabilitation programs. Planning and management of mine action in the region should be strictly in the hands of local people. A centrally nominated and financed authority is certainly not trustworthy to lead the clearing actions.

11. Human Security and Political Impacts of Landmines

Landmines and UXOs cause a serious threat to development and reconstruction of contaminated societies. They have strong negative effects on the process of societies' reintegration in cases suffering from a situation of political instability. The Kurdistan region was not excluded from the phenomenon due to continuous conflict and wars in recent decades which extended to most of the communities in the region. The intensity of contamination differs though from one community to another depending on their strategic location. The civilians are the most frequent victims of war and even in peacetime, many of whom have no choice but to go into dangerous mined areas in order to access basic resources to satisfy their daily lives.

Injuries caused by mines affect the lives of the victims and also impact on their families and the medical infrastructure of the community. The impact extends far beyond individual accidents and casualties and denies access to productive land areas. In sum, landmines and UXOs cause massive collateral damage to a society through their indiscriminate nature and persistent threat long after termination of conflicts.

The concept of human rights covers the political, civil and economic aspects, while the concept of human security focuses on the viewpoints of individuals to protect them from threats, livelihoods and dignity and to utilize their full potential. It is essential to achieve national stability by reducing political tensions. The economic security aspect requires a basic guaranteed work income provided to individuals, or as a last resort of the safety net. The economic impact of landmines can be seen as one of the determinant factors of economic security as it prevents people in the affected area from working and means that the surviving, disabled victims face difficulties in finding employment, and thus remain dependent. In remote areas where the main activities are farming or shepherding, provision of suitable employment to the disabled victims is limited.

Food security ensures that individuals have access to basic physical and economic foodstuffs. Landmines and UXOs have strong impact on food security by making arable land unsafe for cultivation. Thus, the mine clearance has the objective of ensuring that agricultural land returns to productive use. It enables the communities to become self-sufficient and vulnerable people to plant their crops, rebuild their homes and raise their families in a safe manner. In addition, they produce food supplies for the urban population and enhance food security and self-sufficiency. Health security ensures a minimum level of provision of basic health services and protection from disease and unhealthy lifestyles. Landmines and UXOs impact on health security as they kill and maim innocent civilians long after the end of conflicts and block vital areas and infrastructures.

Environmental security aims to protect people from threats in nature and the deterioration of their natural environment. Lack of access to clean water is one of the greatest environmental threats. Landmines and UXOs cause land, soil and habitat degradations, reduce vital resources, alter the natural food chain, and limit agricultural production and productivity. Personal security aims to protect people from physical violence and crime exercised especially by the victims of landmines. Landmines make people risk their safety to generate income for their urgent survival and daily life. In addition, the consequences of being a landmine victim and being disabled will cause these victims psychological problems with undesirable consequences for their surroundings.

Community security aims to protect people from loss of traditional values and relations of violence. Landmines and UXOs often threaten traditional communities, especially targeting ethnic minorities. The impact on communities can be considered an extended level to the impact on individuals, their families assets to reconstruction and rehabilitation programs in the community.

In general, the injuries from landmines will drastically impact the emotional well-being of societies. Human rights violations are most common during periods of political turmoil. Along with the suppression of individuals and groups, governments may try to exert control over the ideas and information flows in the society.

12. Mine Action Resources and Progress

Resources refer to the mine action organizations' capacity, responsibility and resource availability. In recent years, the mine action organization has witnessed major changes. After political agreement on a federal Iraq, the responsibility of mine action was transferred from international organizations and local agencies to the central government. From the Kurds' perspective, this was a major step backwards. This is evidence of lack of awareness and interest of the Kurdistan regional government to the mine issue. This action seems not to be appropriate considering the Iraqi government's crimes against Kurds including the systematic laying of mines in the Kurdistan region. The collected information might be misused by the Iraqi army and security forces in a future conflict. The regional government should actively promote local capacity and force the central government to allocate a special budget to clean up their remains and to rehabilitate the victims of their crimes.

The Kurdistan region is fatally contaminated by mines and UXOs due to different armed conflicts in different time periods. Landmines and UXOs have caused more than 12,800 victims in the region since the early 1970s in which more than 10,500 victims were killed. Landmines constitute a substantial barrier to rehabilitation and development activities in many areas of the region. The resources available to mine action need to be managed effectively to respond to the regional humanitarian and development priorities. The extent of the contaminated area is significantly underestimated due to changed demography and rapid urbanization.

Depopulation of rural areas has led to the fact that not all contaminated areas have been discovered. There is a need to conduct a cost-benefit analysis of mine action activities.

The impacts of landmine contamination and its related accidents in general can be categorized in terms of socio-economic effects of three types: impact on human life, blockage of access to necessary resources for human production and consumption, and the distortion of attitude for various reasons. There is a need to conduct a cost-benefit analysis involving four steps: to identify all the relevant costs and benefits, to place monetary values on these costs and benefits, to convert the stream of future costs and benefits into discounted present values, and calculate net benefit by which a program can be judged. Landmine clearance has numerous benefits, some measurable quantitatively or monetarily and others not. At least three main benefits of mine action are observed in relation with saved lives, medical costs and saved agricultural output from land that otherwise would not be used as a result of mines.

The outcome of mine action in Kurdistan can be improved by using more resources, better techniques and by taking advantage of experience both locally and internationally. However, mine action suffers from a lack of clear vision and strategy. It needs to make a clear feasibility study to show a long-term vision in which it should have important targets. The agency has different financial sources including those from the regional government. However, the local resource allocation is insufficient to correspond to the mine action needs. Very likely, the flow of external resources will be reduced in the near future. As part of a business driven development strategy, the regional government has so far given the highest priority for the construction of different public development infrastructure projects. The agricultural sector has been neglected as part of the reconstruction and development programs in the region.

13. Mine Risk Education and Preventive Measures

Mine action refers to a number of activities which are aimed at reducing the social, economic and environmental impacts of landmines and unexploded ordnances. It is not just about demining. The objective of mine action is thus to reduce the landmine risk to such a level that people can live safely from the constraints imposed by landmines. The mine risk education program is the essential means of reducing the number of deaths and injuries. It aims to teach local populations in contaminated areas to avoid entry to the affected and suspected areas, and thereby effectively reduce the landmine accident rate. Here, the emphasis is on reduction of the effects of landmines on health care and the agricultural production situation, as well as normal human rights in mobility and its negative effects on the families' mental and physical health.

Different programs and campaigns that are based on sharing information, teaching and methods of avoiding mine risks are employed in clearing and accident prevention activities. The resources used include posters and pamphlets. The design and planning of the media component of the action is made by accounting for the multidimensional aspect of the mine problem. The mine awareness program cycle includes a feasibility study, requirement assessment, program planning, monitoring and evaluation steps are required to provide an efficient mine awareness education. The feasibility study helps to identify the needs for mine awareness programs in specific regions. The requirement assessment involves collecting and analyzing relevant data on the population at risk to identify the most appropriate strategy for the mine awareness program. Concerning communication and mine risk education, it should include interpersonal, different media channels, community liaison, and education communication.

14. Legal Actions against Landmine Users and Suppliers

The inequality in use and overuse of some types of conventional weapons has been the main features of the armed conflicts in the current era. It is important to reconsider the related principles of international humanitarian law. It is no longer sufficient to refer only to these principles without finding a mechanism to curb the violations that characterized the contemporary armed conflicts in the region with serious and frequent violation of the above principles. The above principles refer only to use of landmines but not to their production and supply.

Two important principles of international humanitarian law related to the use of certain conventional weapons are addressed, namely the military necessity and the proportionality and adaptation of the use of such weapons. In this regard, the landmine and cluster munitions on the grounds are considered the most hideous and barbarous weapons. It is clear that the international humanitarian laws are not effective or sufficient. Countries and non-international parties in armed conflicts are often trying to justify the serious violation based on the principle of military necessity and the proportionality. Since there is no need for determination of the legitimacy of using certain types of such weapons, the conflict parties continue to violate the international humanitarian law.

In order to curb these serious violations, there must be a specific mechanism to investigate them and to prosecute those responsible to deter producers and suppliers of such inhuman weapons. The countries responsible to the formation of such a court should be the same countries who led the negotiating track in the two humanitarian agreements, namely the Ottawa Convention to ban landmines of 1997 and the Dublin Convention of 2008 to ban cluster munitions. In the process, they should take advantage of the experience and evidence of atrocities in the countries and regions like Kurdistan.

The major powers with access to technical, financial, and organizational capacity should contribute to the formation and operation of such noble activity. However, the countries with veto power in international institutions seem not to be legitimate partners. These have shown in practice that with no exception they take sides with their businesses and strategic interests. Under such conditions and strong influence it seems that improvements along the lines discussed above which would have favoured the suffering civilian populations inhabiting countries engaged in war are not forthcoming.

Considering the Kurdistan Region, we suggest an institute be established to conduct research on landmine and analyze collected statistics. Such an institute will play a major role in the coordination of research related to different aspects of landmines such as information, prevention, training, disability, rehabilitation and the labour market. In addition, it will enable the gathering of evidence on the crimes committed, supporting the regional government in its efforts to follow up the landmine issue for matters of prevention, reducing impacts, victim compensation, and bringing the landmine users to justice.

APPENDIX A

METHODOLOGY APPLIED FOR CASUALTY ANALYSIS

1. Victim's Basic Statistics

Statistics

Recent victims

N	Valid	1021
	Missing	0
Mean		.31
Median		.00
Mode		0
Std. Deviation		1.030
Skewness		6.600
Std. Error of Skewness		.077
Kurtosis		61.622
Std. Error of Kurtosis		.153
Minimum		0
Maximum		13
Sum		313
Percentiles	25	.00
	50	.00
	75	.00

2. Independent Variables' Basic Descriptive Statistics

N	Vali		Number	Distance neare othe communi with rece victim (log1	Distance internation bord (log1	Populatio (number
	Missin	Has 0	mine 0	meter 0	mete 0	familie 2
Mea		.35	3.2	3.743	4.347	144.6
Media		.00	2.0	3.743	4.301	25.0
Mod		0	1	2.8 a	4.3	0
Std.		.47	3.08	.3726	.4414	755.86
Skewne		.65	2.87	-	-	11.00
Std. Error of		.07	.07	.07	.07	.07
Kurtos		-	12.06	.15	.70	149.27
Std. Error of		.15	.15	.15	.15	.15
Minimu		0	1	2.2	2.8	0
Maximu		1	29	4.8	5.2	1371
Percentil	25	.00	1.0	3.502	4.000	10.0
	50	.00	2.0	3.743	4.301	25.0
	75	1.0	4.0	4.014	4.602	54.0
a						

3. Zero-Inflated Negative Binomial Regression

We used this model, instead of zero-inflated Poisson, because of significant overdispersion. The STATA command for the final model was:

zinb victimsr hasuxo nominefi log10distNOCVRV log10distborder log10pop, inflate(hasuxo nominefi log10distNOCVRV log10distborder log10pop) robust, and the output is as follows:

Zero-inflated negative binomial regression Number of obs = 958
Nonzero obs = 160
Zero obs = 798

Inflation model = logit Wald chi2(5) = 69.08
Log likelihood = -554.9712 Prob > chi2 = 0.0000

Robust standard errors

Variable	Coefficient	Standard Error	z-value	P>|z	[95% Conf. Interval]	
victimsr						
hasuxo	0.679491	0.214132	3.17	0	0.259799	1.099182
nominefi	0.030194	0.019963	1.51	0.13	-0.008932	0.069322
log10distN~V	-1.119362	0.340528	-3.29	0	-1.786786	-0.45194
log10distb~r	0.134631	0.310317	0.43	0.66	-0.473579	0.742841
log10pop	1.001733	0.162008	6.18	0	0.684202	1.319264
_cons	0.374404	1.5006	0.25	0.8	-2.566718	3.315527
inflate						
hasuxo	0.666521	1.080766	0.62	0.54	-1.451741	2.784783
nominefi	-2.517439	1.534437	-1.64	0.1	-5.52488	0.490001
log10distN~V	-0.37672	2.78675	-0.14	0.89	-5.838649	5.085209
log10distb~r	4.293619	4.534811	0.95	0.34	-4.594448	13.18169
log10pop	-1.993902	0.933271	-2.14	0.03	-3.823079	-0.16472
_cons	-12.09711	13.87946	-0.87	0.38	-39.30036	15.10615
/lnalpha	0.600184	0.127482	4.71	0	0.350322	0.850045
alpha	1.822454	0.232331			1.419525	2.339753

spearman victimsr victimsrpredict

Number of obs = 958
Spearman's rho = 0.3520

Test of Ho: victimsr and victimsrpredict are independent, Prob > |t| = 0.0000.

APPENDIX B

UNOPS MAP FIELD MODULE

Main Tables

(1) **GDSV All Villages**: The "GDSV" is an acronym deriving from the words "Governorate," "District," "Sub District" and "Village." The table contains the structure of Northern IRAQ Villages and its affiliation to the Sub-Districts, Districts, and Governorate.
(2) **Main Villages**: This table collects information about the area affected by mines/UXOs and its social-economic impact on the society living in the immediate surroundings.
(3) **SUSPECTED MINED AREA**: Recording all information about suspected mined area, the records in this table contain minefield data as well as some areas that are not minefields but are affected by mines.
(4) **Main MINEFIELDS**: This table records general survey data for all the minefields, and high impact survey data after re-verifying it by survey teams. This table is a subset of the "SUSPECTED MINED AREA" table.
(5) **Main MINEFIELDS Completion Report:** This is a record of a completed UXO or mine clearance task, which contains all the survey and clearance final data as well as the actual size and boundary of cleared land.
(6) **Main MINEFIELDS Task Orders**: Stored data of the technical survey after doing the pilot visit for a specific minefield.
(7) **Main MINEFIELDS Site Design**: This table records the site design visits to the minefield by the technical survey teams to determine the boundary of the minefield and to divide the area for resources.
(8) **Main MINEFIELDS BOI**: Stored data of the board of inquiry, the board that will be convened by UNOPS Project Coordinator to investigate demining accidents and for any other incident that the project coordinator deems necessary.
(9) **Main PIA:** Stored data that determines the impact of the cleared land on the owners and community after six months from handing over.

(10) **VR Main Verification:** This table records information about cleared minefields by villagers, MAG, Solidarity, NPA and other demining committees. Survey teams visit those areas and bring back verified data to be stored in the database.

(11) **Main INCIDENTS:** Stored data of mine/UXO incidents/ accidents from the general survey process.

(12) **Main Recent:** Stored data of recent mine/UXO victims from the general survey process.

Appendix B1 IMSMA Impact Survey

IMSMA

Impact Survey

Cover Sheet Locator code: ...../...../...../...

1 General

	Name	Code	
1.1 Country			1.10 **Confirmed:** ☐ Yes ☐ No
1.2 Province			1.11 Reliability of information: ☐ 1 ☐ 2 ☐ 3 ☐ 4 ☐ 5 ☐ 6
1.3 District			1.12 Reliability of source: ☐ A ☐ B ☐ C ☐ D ☐ E ☐ F
1.4 Subdistrict			1.13 Survey ID: (Leave blank)
1.5 Data enumerator organisation			1.14 Owner MAC:
1.6 Data enumerator 1			1.15 Referenced TaskID(s):
1.7 Data enumerator 2			1.16 Survey planned start date:
1.8 Area officer			1.17 Survey planned completion date:
1.9 Area supervisor			1.19 Survey effective start date:
1.18 Status			1.20 Survey effective completion date:

2 Location:

2.1 Town name: __________

2.2 Town code: __________

2.3 Name alternative 1: __________

2.4 Name alternative 2: __________

2.5 Name alternative 3: __________

2.6 Name alternative 4: __________

2.7 Name alternative 5: __________

2.8 Municipality: __________

3 Geographical reference:

3.1 Coord. system:	3.6 Map name:
3.2 X/ Easting/ Long:	3.6 Map series:
3.3 Y/ Northing/ Lat:	3.6 Map edition:
3.4 MGRS Coord[1]:	3.6 Map sheet:
3.5 Coordinate fixed by: ☐ DGPS ☐ GPS or Map with accuracy: ☐ <30m ☐ >30 m	3.6 Map scale: **1 :**

3.7 Description of the geographical reference:

[1] MGRS provided when X/Y absent and vice versa.

Created by IMSMA v. Appendix B1 IMSMA Imcpact Survey.doc (Rev. 17.01.03)

IMSMA

Impact Survey
Interview Group

[4]Interview group data

	Name[2]		4.3 Sex		4.4 Age[3]					4.5 Mine	4.6 Vocation[4]				
	4.1 Family name	4.2 First name	M	F	5 - 14	15 - 29	30-44	45-59	60 +	victim	1	2	3	4	5
1			☐	☐	☐	☐	☐	☐	☐	☐	☐	☐	☐	☐	☐
2			☐	☐	☐	☐	☐	☐	☐	☐	☐	☐	☐	☐	☐
3			☐	☐	☐	☐	☐	☐	☐	☐	☐	☐	☐	☐	☐
4			☐	☐	☐	☐	☐	☐	☐	☐	☐	☐	☐	☐	☐
5			☐	☐	☐	☐	☐	☐	☐	☐	☐	☐	☐	☐	☐
6			☐	☐	☐	☐	☐	☐	☐	☐	☐	☐	☐	☐	☐
7			☐	☐	☐	☐	☐	☐	☐	☐	☐	☐	☐	☐	☐
8			☐	☐	☐	☐	☐	☐	☐	☐	☐	☐	☐	☐	☐
9			☐	☐	☐	☐	☐	☐	☐	☐	☐	☐	☐	☐	☐
10			☐	☐	☐	☐	☐	☐	☐	☐	☐	☐	☐	☐	☐
11			☐	☐	☐	☐	☐	☐	☐	☐	☐	☐	☐	☐	☐
12			☐	☐	☐	☐	☐	☐	☐	☐	☐	☐	☐	☐	☐
13			☐	☐	☐	☐	☐	☐	☐	☐	☐	☐	☐	☐	☐
14			☐	☐	☐	☐	☐	☐	☐	☐	☐	☐	☐	☐	☐

4.7 In which language was the group interview primarily conducted? _____

[2] Interviewee name should be entered with English (Latin character set).
[3] Age categories derived from the Cambodia Red Cross Victims surveillance program.
[4] Five categories such as: farmer, merchant, soldier or housewife. To be determined by the country MAC

IMSMA

Impact Survey

Locality Locator code: .../.../.../...

[5] **External / Expert opinion (prior to Impact Survey):**

[5.1] Origin of impact survey report: ☐ National Capital ☐ Provincial Headquarters ☐ District Headquarters ☐ While visiting another community ☐ Former non-affected community ☐ Other ______________

[5.2] Kind of expert responsible: ☐ Government ☐ NGO ☐ UN ☐ Member of Community ☐ Other ______________

[5.3] Judgment of area condition prior to visiting: ☐ Affected ☐ Possibly affected ☐ Not affected

[6] **Historical / Conflict information:**[5]

[6.1] Last conflict in the area: _____ (year)

[6.2] Year mines problem began: _____ (year)

[6.3] Year mines last planted: _____ (year)

[6.7] Comments on local history of mines problem

[6.8] Intensity of military activity:

[6.9] When was settlement established?: ______________________________ (year or description)

[7] **Mine Risk Education:**[6]

[7.1] Mine risk education took place within past 24 months: ☐ Yes ☐ No ☐ Unknown

Select:

☐	School presentation	☐ Last 24 months	☐ over 24 months ago
☐	Posters	☐ Last 24 months	☐ over 24 months ago
☐	Radio	☐ Last 24 months	☐ over 24 months ago
☐	Performing group	☐ Last 24 months	☐ over 24 months ago
☐	Community center	☐ Last 24 months	☐ over 24 months ago
☐	Religious organisation	☐ Last 24 months	☐ over 24 months ago
☐	Other:	☐ Last 24 months	☐ over 24 months ago

[6.2] Organisations involved in recent mine risk education (if any):

[5] Date (Year) that last conflict took place in the area should be entered.

[6] Mine risk education given during the specified time frame?

Created by IMSMA v. Appendix B1 IMSMA Imcpact Survey.doc (Rev. 17.01.03)

IMSMA

Impact Survey

Locality Locator code: .../.../.../...

[7] Victim totals

	[7.1] Number killed	[7.2] Number injured	[7.3] Total
Number of recent victims			
Number of victims more than 24 months ago and recent victims not specified [7]			
Total number of victims			

[8] Victim assistance during past 24 months: ☐ Yes ☐ No ☐ Unknown

[9] Locality:

	[9.1] Pre-war population		[9.2] Current population	
Data collected	Census	Key informants	Census	Key informants
Number of households				
Total population				

[9.3] Marking and survey during the past 24 months: ☐ Yes ☐ No ☐ Unknown

[9.4] Mine clearance during the past 24 months: ☐ Yes ☐ No ☐ Unknown

[9.5] Local clearance efforts: ☐ Yes ☐ No ☐ Unknown

If yes, effects of local clearance efforts: _____

[9.6] Piped water supply: ☐ Yes ☐ No ☐ Unknown

[9.7] Electricity supply: ☐ Yes ☐ No ☐ Unknown

[9.8] Fuel available: ☐ Yes ☐ No ☐ Unknown

[9.9] Telephone connection: ☐ Yes ☐ No ☐ Unknown

[9.10] Medical facility: ☐ Basic health facility ☐ Hospital ☐ None ☐ Other ☐ Unknown

[9.11] Number of primary schools: _____

[9.12] Number of secondary schools: _____

[9.13] Presence of higher education institutions: ☐ Yes ☐ No ☐ Unknown

If yes, what institutions? _____

[9.14] Economic base: ☐ Agriculture/ husbandry ☐ Industry ☐ Tourism ☐ Government ☐ Other

What other? _____

Impact Survey
Mined Area Locator code: .../.../.../...

(One sheet per mined area)

10 Mined area

10.1 ID (consecutively)	10.2 Name of mined area (geographic location)

Viewing point

10.4 Starting point measured from viewing point ☐ Yes ☐ No

Geographic coordinates of viewing point

10.5 Coordinate system:

10.7 Map east	10.8 Map north	10.9 MGRS[8]

Starting point of mined area (from viewing point or from geographical reference of community)

10.11 Bearing degrees[9]	Distance	
	10.12 Meters[10]	10.13 Walk time (min)

Starting point converted to geographic / UTM / MGRS coordinates

10.14 X/ Easting/ Longitude	10.15 Y/ Northing/ Latitude	10.16 MGRS

10.17 Coordinates visually verified ☐ Yes ☐ No

10.18 Description of the mined area and starting point:

10.20 Estimated size of mined area: _____ m^2

10.21 Marking (choose only one): ☐ Local signs ☐ Official signs ☐ Fenced ☐ None

10.22 Was there fighting? ☐ Yes ☐ No ☐ Unknown

10.23 Status of mined area: _____

[8] MGRS provided when X/Y absent and vice versa.
[9] Bearing in degrees, geographic. North is 0 degrees and 360 degrees; east is 90 degrees; south is 180 degrees; and west is 270 degrees.
[10] Estimated distance in meters from the town center coordinate to the closest point of mined area.

IMSMA

Impact Survey

Mined Area Locator code: .../.../.../...

[11] **Suspected ordnance in the area:**

[11.1] Type of Device	[11.2] Model (description)
☐ AP ☐ AT ☐ UXO ☐ Unknown	
☐ AP ☐ AT ☐ UXO ☐ Unknown	
☐ AP ☐ AT ☐ UXO ☐ Unknown	
☐ AP ☐ AT ☐ UXO ☐ Unknown	
☐ AP ☐ AT ☐ UXO ☐ Unknown	
☐ AP ☐ AT ☐ UXO ☐ Unknown	

IMSMA

Impact Survey

Mined Area Locator code: .../.../.../...

(One sheet per mined area)

[12.1] **Agricultural fields blocked:**

Select: ☐ Crop type: ⇒ ☐ Irrigated ☐ Rain fed ⇒ ☐ Grain ☐ Fruit ☐ Vegetable ☐ Other ☐ Unknown

☐ Pasture Type: ⇒ ☐ Fixed pasture ☐ Migratory pasture ⇒ ☐ Cattle ☐ Goats, sheep, pigs ☐ Other ☐ Unknown

[12.2] **Non-agricultural areas blocked:**

Select: ☐ Fuel
☐ Food
☐ Building materials
☐ Medicinal

[12.3] **Water access blocked:**

Select: ☐ Irrigation
☐ Fishing
☐ Watering animals
☐ Bathing
☐ Laundry
☐ Other
☐ Drinking: ⇒ ☐ Lake, Stream etc. ☐ Well, Spring, etc.

[13.1] **Infrastructure blocked:**

☐ Housing area blocked

Roads blocked to:

☐ District center ☐ Other: _____
☐ Provincial capital ☐ Alternative routes: _____
☐ National capital

[13.2] **Other infrastructure blocked (one or more):**

☐ Bridge ☐ Factory
☐ Dam or canal ☐ Oil field
☐ Railroad ☐ Medical facility
☐ Airstrip ☐ Educational facility
☐ Power line ☐ Market
☐ Power station ☐ Cultural site
☐ Other vital points: What: _____

[13.3] Is a development project planed in the area: ☐ Yes ☐ No ☐ Unknown

[13.4] If yes, is the project funded: ☐ Yes ☐ No ☐ Unknown

[13.5] Contact info for project: _____

IMSMA

Impact Survey

Mined Area Locator code: .../.../.../...

(One sheet per mined area)

14 Terrain data

14.1 Vegetation: ☐ Short grass ☐ Tall grass ☐ Bushes ☐ Trees ☐ None ☐ Unknown ☐ Other: ____

14.2 Ground profile: ☐ Flat ☐ Hillside ☐ Ridge ☐ Gully ☐ Unknown ☐ Other: ____

14.3 Special features:

14.4 Major impacts of this mined area:

Person that identified this mined area

14.5 Name: ____

14.6 Contact information:

Mined area sketch

14.7 A ttach explanatory map or sketch:

14.8 Picture ID's:

Impact Survey

Mined Area Locator code: .../.../.../...

(One sheet per individual victim)

15 Recent Victims

15.1 Victim number:[11] ☐

15.2 Known dangerous area: ☐ Yes ☐ No

15.3 Date of detonation [if known]:[12] ______ (year / month / day)

Name [if known]

15.4 Family name:	
15.5 1st name:	
15.6 2nd name:	

15.7 Sex and 15.8 age:

☐ Male ☐ Female ☐ Unknown

☐ 0 – 4 year old
☐ 5 - 14 years old
☐ 15 - 29 years old
☐ 30 – 44 years old
☐ 45 - 59 years old
☐ 60 years and older

15.9 Occupation:

☐ Artisan
☐ Farming
☐ Herding
☐ Household work
☐ Military
☐ Office work
☐ Trading
☐ Was not earning
☐ Other What other:
☐ Unknown

15.10 Activity:

☐ Military
☐ Civilian
- ☐ Travel
- ☐ Tampering
- ☐ Farming
- ☐ Household work
- ☐ Collecting food, water, etc.
- ☐ Herding
- ☐ Playing
- ☐ Unknown
- ☐ Other

15.11 Wounds:

☐ Fatal
☐ Amputation
☐ Loss of Sight
☐ Other body wound
☐ Unknown

15.12 Care:

☐ Emergency medical
☐ Physical rehabilitation
☐ Vocational training
☐ Other
☐ None
☐ Immediately fatal
☐ Unknown

---

[11] The victim number is assigned in a sequential order by the enumerator.

[12] This sheet is to be completed for each victim/survivor during the last 24 months. The time and detonation is being asked here as a check that the incident happened during the last 24 months. Date should be entered YYYY-MM-DD.

IMSMA

Impact Survey
Contacts

Locator code: .../.../.../...

[16] Local contacts[13]:			
Name[14]	Address[15]	Communications[16]	Remarks

[13] Not only contacts collected during survey, but all contacts from town.
[14] Fill in name, first name, and position.
[15] Fill in organisation, department, address, postal code, town, and country.
[16] Fill in phone and fax numbers, e-mail address, call sign, frequency, and channel.

IMSMA

Impact Survey

General Comments / Data Collector Information / Attachment Locator code: .../

17.1 **General comments:**

Impact adjustment:

17.2 Correction factor: ☐ -20% ☐ -10% ☐ 0 ☐ +10% ☐ +20%

17.3 Justification of correction factor:

17.4 Major mine impacts:

17.5 Change of severity: ☐ Less severe ☐ Same ☐ More severe ☐ Not applicable ☐ Unknow

18 **Status History**

Status	Valid from	Status user	Reason	Comment

19 **Data collection information:**

19.1 Tasked by	
19.2 Date report received	

19.5 Area supervisor	
Date & signature	

19.3 Data entered by	
19.4 Date & signature	

19.6 Verified by	
19.7 Date of verification	

19.8 Enumerator name 1	
19.9 Enumerator name 2	
19.10 Date & signature	

Total number of pages	11

20 **Community leader certifying:**

20.1 Name	
20.2 Position	
Date & signature	

21 **Attachment:**

- General location & Contacts
- Sketch:

Appendix B2 IMSMA Minefield Survey

IMSMA **Minefield Report** Locator code: .../.../.../...

[1] General Mine Field Information

		[1.11] **Confirmed:**	☐ Yes ☐ No
[1.3] Technical survey ID:			
[1.4] Minefield name:		[1.14] Reported by:	
[1.5] Referenced TaskID(s):		[1.15] Position:	
[1.6] Province:		[1.16] Organisation:	
[1.7] District:			
[1.8] Subdistrict:		[1.18] Date of report:	
[1.9] Nearest town:			

[1.20] Distance from nearest town: _____ km

[1.21] Road possible for: ☐ Small 4x4 ☐ Big truck ☐ All

[1.22] Best road to minefield from nearest town:

[1.23] Minefield description:

[2] Medical facilities:

	Name of location [specify details in "location sheet" for level 3,4, med.]	Travel time in hours & min
Level 1:	2.1	2.2
Level 2:	2.3	2.4
Level 3:	2.5	2.6
Level 4:	2.7	2.8
Medivac:	2.9	2.10

IMSMA **Minefield Report** Locator code: .../.../.../...

3 Minefield geographical reference and perimeters

3.1 Coord. system:	3.2 Map edition:
3.2 Map name:	3.2 Map sheet:
3.2 Map series:	3.2 Map scale: 1:
5 Coordinates of the landmark:	**6 Coordinates of the benchmark:**
5.1 X/ Easting/ Long:	6.1 X/Easting/Long:
5.2 Y/ Northing/ Lat:	6.2 Y/Northing/Lat:
5.3 MGRS Coord:	6.3 MGRS Coord:
5.4 Coordinates fixed by: ☐ DGPS ☐ GPS or Map with accuracy: ☐ <30m ☐ >30m	6.4 Coordinates fixed by: ☐ DGPS ☐ GPS or Map with accuracy: ☐ <30m ☐ >30m
5.5 Description of landmark:	6.5 Description of benchmark:

6 Description of the minefield perimeter points

From point	No.	7.1 To point	7.2 Bear. [°]	7.3 Dist. [m]	7.4 X/ East/ Long.	7.5 Y/ North/ Lat.	7.6 MGRS Coord.
Benchmark	1	Datum point					
Datum point	2						

8.1 Size of area: _____ m^2

8.2 Marking: ☐ Local signs ☐ Official signs ☐ Fenced ☐ None

IMSMA **Minefield Report** Locator code: .../.../.../...

[7] Estimated Devices

Type of Device (AP, AT, UXO, etc.)	Model	Estimated Quantity

[7.1] Estimated devices: Total: _____

[7.2] Information Classification:

	Code	Evaluation	Information	Source
☐	M1	Mines or UXO physically verified	Confirmed	Reliable
☐	M2	Area reported with observed evidence of mines or UXO	Unconfirmed	Reliable
☐	M3	Area reported with observed evidence of mines or UXO	Unconfirmed	Unreliable
☐	M4	Area reported as mined with no evidence or indications of mines or UXO	Unconfirmed	Unreliable

[7.3] Condition of mines/UXO: ☐ Unfused ☐ Safe ☐ Blind ☐ Unknown

[8] Demining (operational information)

[8.01] Vegetation: ☐ Grass ☐ Bushes ☐ Trees ☐ None

[8.02] Vegetation density: ☐ Low ☐ Medium ☐ High

[8.03] Vegetation removable by: ☐ Manual ☐ Mechanical ☐ Chemical ☐ Burning

[8.04] Terrain category:
☐ A (open, week ground)
☐ B (hard ground mixed with rocks and light vegetation)
☐ C (very hard stony ground with heavy vegetation)
☐ D (wet/mud)

[8.05] Soil Type: ☐ Sand ☐ Chalk ☐ Ploughed ☐ Clay ☐ Rocky ☐ Swamp

[8.06] Contamination with: ☐ Metal ☐ Rubble ☐ Tree stumps ☐ Other _____

[8.07] Level of contamination with metal: ☐ Low ☐ Medium ☐ High

[8.08] Level of soil metal content: ☐ Low ☐ Medium ☐ High

[8.09] Drainage features: ☐ Canal ☐ River ☐ Lake ☐ Other _____

[8.10] Ground profile: ☐ Flat ☐ Hillside ☐ Ridge ☐ Gully

[8.11] Slope: ☐ 0-5% ☐ 5-10% ☐ 10-15% ☐ 15-20% ☐ 20-25% ☐ 25-30% ☐ 30% +

[8.12] Is the terrain suitable to use dogs? ☐ Yes ☐ No ☐ Unknown

[8.13] Is the terrain suitable to use detectors? ☐ Yes ☐ No ☐ Unknown

[8.14] Is the terrain suitable to use prodding? ☐ Yes ☐ No ☐ Unknown

IMSMA **Minefield Report** Locator code: .../.../.../...

8.15 Is mechanical demining possible? ☐ Yes ☐ No ☐ Unknown

8.16 What type of mechanical device can be brought to the minefield? ☐ Light ☐ Medium ☐ Heavy

8.17 Special seasonal consideration:

9 Recommendation

9.1 Assessed degree of difficulty: ☐ 1 ☐ 2 ☐ 3 ☐ 4 ☐ 5

9.2 Recommended clearance priority: ☐ Low ☐ Medium ☐ High

9.3 Recommended no. of demining teams: _____

9.4 Type of operation recommended: ☐ Manual ☐ Mechanical ☐ Dogs ☐ Combined

9.5 Depth of clearance recommended: _____cm

9.6 Reason for clearance depth:

9.7 Estimated time for completion of clearance: _____days

9.8 Intended land use: ☐ Housing ☐ Development ☐ Industrial ☐ Agricultural ☐ Community ☐ Other: _____

9.9 Date of intended land use: _____ (month/year)

10 Agricultural fields blocked:

Select: ☐ Crop type: ⇒ ☐ Irrigated ☐ Rain fed ⇒ ☐ Grain ☐ Fruit ☐ Vegetable ☐ Other ☐ Unknown

☐ Pasture Type: ⇒ ☐ Fixed pasture ☐ Migratory pasture ⇒ ☐ Cattle ☐ Goats, sheep, pigs ☐ Other ☐ Unknown

IMSMA **Minefield Report** Locator code: .../.../.../...

[11] **Water access blocked:**

Select:
- ☐ Irrigation
- ☐ Fishing
- ☐ Watering animals
- ☐ Bathing
- ☐ Laundry
- ☐ Other
- ☐ Drinking: ⇒ ☐ Lake or stream
  ☐ Well or spring

[12] **Non-agricultural areas blocked:**

Select:
- ☐ Fuel
- ☐ Food
- ☐ Building materials
- ☐ Medicinal

[13] **Infrastructure blocked:**

- ☐ Housing area blocked

Roads blocked to:

- ☐ District center ☐ Other: ____
- ☐ Provincial capital ☐ Alternative routes: ____
- ☐ National capital

[14] **Other infrastructure blocked:**

- ☐ Bridge
- ☐ Dam or canal
- ☐ Railroad
- ☐ Airstrip
- ☐ Power line
- ☐ Power station
- ☐ Other vital points: What: ____
- ☐ Factory
- ☐ Oil field
- ☐ Medical facility
- ☐ Educational facility
- ☐ Market
- ☐ Cultural site

[15] Is a development project planed in the area? ☐ Yes ☐ No ☐ Unknown

[15.1] If yes, is the project funded? ☐ Yes ☐ No ☐ Unknown

[15.2] Contact info for project: ____

[16] Mine risk education given: (How trained/who trained)

(use MRE activity instead of this field)

IMSMA **Minefield Report** Locator code: .../.../.../...

[17] Additional information/comments

[19] Data source information:

[19.1] Reported (Collected) by:	
Location:	
Date:	
Signature:	

[19.2] Verified by:	
Location:	
[19.1] Date:	
Signature:	

[19.3] Data entry by:	
[19.4] Date entry date:	
Signature	

Total number of pages: 7

Attachments:

Attach a sketch map

Map of detailed area showing mined area, area to be cleared, access roads, recommanded campsite and emergency landing zones, local landmarks etc. Include a north arrow, scale and legend.

IMSMA

Minefield Report

Detailed sketch of area(s) with mine/UXO contamination Locator code: .../.../.../...

1 Square(5mm) = meters

N

CHECKLIST

- Landmark
- Benchmark
- Datum point
- Turning points
- Start line
- Safe lanes
- Safe areas
- Cleared access roads
- Mine rows
- Location of booby traps
- Legend

Drawn by: **Date:** **Checked by:** **Date:**

Appendix B3 IMSMA Victim Survey

IMSMA **Victim** Locator code: ../../../...

[2] Victim data

[2.1] Victim ID:	[2.2] Owner MAC:	
[2.3] Family name:	[2.5] Sex: ☐ Male ☐ Female	[2.7] Address:
[2.4] First name:	[2.6] Date of Birth:	

[1] General mine accident information:

[1.1] Mine accident ID:		[1.6] Data entry date:	
[1.2] Date and time of mine acc:		[1.7] Data entry by:	
[1.3] Data gathered by:		[1.8] Date of report:	
[1.4] Reported by:		[1.9] Date of report received:	
[1.5] Organisation:	(Address & Tel)		

Nearest town from mine accident

[1.10] Province:	[1.12] Subdistrict:
[1.11] District:	[1.13] Nearest town:
	[1.14] Municipality:

Distance and direction from nearest town (Not necessary, if coordinates are known):

[1.20] Distance from nearest town: ☐ Less than 500m ☐ 500 m – 5 km ☐ More than 5 km

[1.21] Direction from nearest town: ☐ North ☐ South ☐ North – East ☐ South – East ☐ East ☐ West ☐ North – West ☐ South - West ☐ Unknown

[3] Injuries:

[3.1] Was the person injured or killed: ☐ Killed ☐ Injured

[3.2] If killed, location of death:
☐ In situ ☐ At health care facility
☐ During transport to health care facility
☐ Other: _____

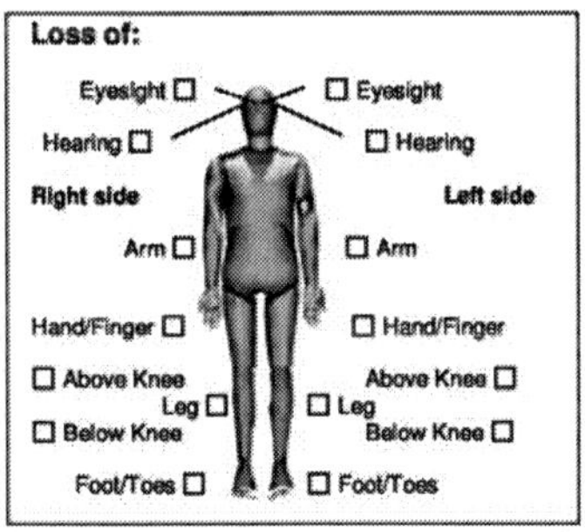

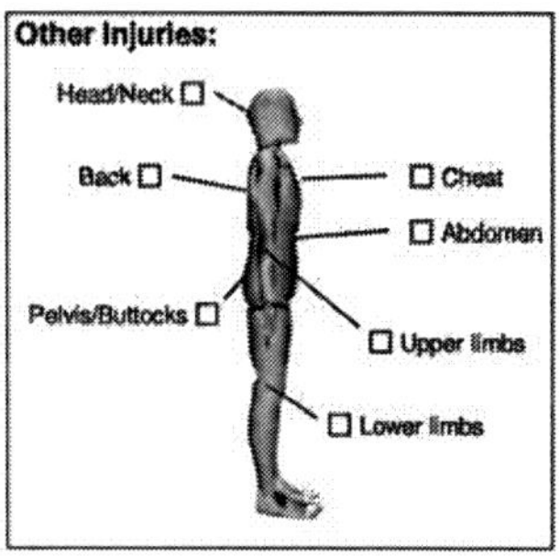

IMSMA **Victim** Locator code: .../.../.../...

[4] Other Information:

[4.1] First medical facility reached: ☐ Dispensary ☐ First aid ☐ Hospital
[4.2] Time until first facility reached: _____h
[4.3] Name of first hospital reached: _____
[4.4] Time until first hospital reached: _____h

[4.13] Occupation:

☐ Mine action personnel	►	☐ Contractor ☐ Government ☐ MAC ☐ NGO ☐ UN
☐ Military	►	☐ Int. peacekeeper ☐ National
☐ Civilian	►	☐ IDP ☐ Local resident ☐ Passing through ☐ Pastoralist/nomad ☐ Refugee
☐ Aid worker ☐ Government official ☐ International observer ☐ Other ☐ Unknown		

[4.14] Occupation *prior* to mine accident

☐ Mine action personnel	►	☐ Contractor ☐ Government ☐ MAC ☐ NGO ☐ UN
☐ Military	►	☐ Int. peacekeeper ☐ National
☐ Civilian	►	☐ IDP ☐ Local resident ☐ Passing through ☐ Pastoralist/nomad ☐ Refugee
☐ Aid worker ☐ Government official ☐ International observer ☐ Other ☐ Unknown		

[4.5] Activity at time of mine accident:

☐ Tending animals/livestock	☐ Passing/standing nearby	☐ Collecting wood/food / water	☐ Hunting/fishing
☐ Demining	☐ Military	☐ Police	☐ Playing/recreation
☐ Tampering	☐ Farming	☐ Travelling	☐ Household work
☐ Unknown	☐ Other: _____		

[4.6] How often did the person go there? ☐ More than once a day ☐ Once a day ☐ Several times a week or less ☐ Never before

[4.7] Did the person know that area was dangerous? ☐ Yes ☐ No ☐ Unknown

[4.8] If they knew area was dangerous, why did they go there? ☐ No other access ☐ Economic necessity ☐ Peer pressure ☐ Other _____

[4.9] Did the person see the object before accident? ☐ No ☐ Yes, did not touch ☐ Yes, touched it ☐ Unknown

[4.10] Did the person receive Mine Risk Education? ☐ Yes ☐ No ☐ Unknown

[4.11] Medical report reference (if available): ☐

[4.12] Was area marked? ☐ Yes ☐ No

IMSMA **Victim** Locator code: .../.../.../...

[5] Received MRE: *Use values from the lists provided in the Appendix!*

[5.1] Activity class	[5.2 – 5.3] Method	[5.4] Frequency

[6] Other persons involved ☐

How many others were killed ? _____

How many others were injured? _____

List of other Victims

[6.2] First name	[6.1] Name	[6.3] Status
		☐ Killed ☐ Injured
		☐ Killed ☐ Injured
		☐ Killed ☐ Injured

[6] Device that caused the mine accident

☐ [2.1] Unknown ☐ [2.2] Anti-personnel mine ☐ [2.3] Anti-tank mine ☐ [2.4] Cluster munitions ☐ [2.5] Other UXO

☐ [2.6] Booby trap ☐ [2.7] Fuse ☐ [2.8] Specify device, if it is known: _____

Appendix

[5.1] Activity class / [5.2] Method:

Community liaison: • Community mapping • Marking maintenance • Minefield handover • Support to demining • [5.3] Other: what other?

Info. management: • Data gathering • Evaluation • Monitoring • [5.3] Other: what other?

Media: • Material distribution • Printed press • Radio • TV • Video • [5.3] Other: what other?

Presentation: • Community • Safety briefing • School • [5.3] Other: what other?

Public performance: • Festival • Music • Poetry • Theater • Traveling road show • [5.3] Other: what other?

Training: • Child-to-child • Peer-to-peer • Lecture/Classroom • Community mobilisation • [5.3] Other: what other?

[5.4] Frequency: • Once • Several times • Regularly

Appendix B4 IMSMA Completion Survey

IMSMA **Completion Survey**

Locator code: .../.../.../...

[1]General Information

[1.1] ID:		[1.7] Data entry date:	
[1.2] Owner MAC:		[1.8] Data entry by:	
[1.3] Referenced TaskID(s):			
[1.4] Organisation:		[1.9] Total size of cleared area:	m²
[1.5] Planned start date:		[1.10] Clearance depth:	cm
[1.6] Planned end date:		[1.11] Total of work hours:	h
[1.5] Task start date:		[1.5] Task end date:	
[1.7] Status:			

[1.12] Object: ☐ Housing ☐ Industry ☐ Agricultural terrain ☐ Other

[1.13] Methods and technologies used? ☐ Manual ☐ Mechanical ☐ Dogs ☐ Combined

[1.14] Based on: ☐ Dangerous area ☐ Clearance ☐ Minefield (technical survey)

[1.15] Reference ID:

[1.16] Accidents occurred during activity: ☐ Yes ☐ No [1.17] Accident IDs:

[1.18] Uncleared area left: ☐ Yes ☐ No

[1.19] Uncleared area marking: ☐ Local signs ☐ Official signs ☐ Fenced ☐ Several ☐ None

[1.20] Description of uncleared area:

[1.21] Additional information:

IMSMA **Completion Survey**

Locator code: .../.../.../...

1.22 Quality control carried out by: ______(name & organisation)

1.23 Quality control method: ☐ Manual ☐ Dogs ☐ Mechanical ☐ Combined

2 Location

2.1 Province:	2.6 Coord. system:	2.11 Map name:
2.2 District:	2.7 X/ Easting/ Long:	2.11 Map series:
2.3 Subdistrict:	2.8 Y/ Northing/ Lat:	2.11 Map edition:
2.4 Nearest city:	2.9 MGRS Coord:	2.11 Map sheet:
2.5 Municipality:	2.10 Coord. fixed by: ☐ DGPS ☐ GPS or Map with accuracy: ☐ <30m ☐ >30 m	2.11 Map scale:1 :

2.12 Description of reference point:

IMSMA **Completion Survey**

Locator code: .../.../.../...

[3] Description of the perimeter points:

From Point	No.	To Point	[3.1] Bearing	[3.2] Distance	[3.4] X/ East/ Long	[3.5] Y/ North/ Lat	[3.6] MGRS Coord.
	1						
	2						
	3						
	4						
	5						
	6						
	7						
	8						
	9						
	10						
	11						
	12						

[3.1] Size of area: _____ m^2

[4] Quantity and type of destroyed devices:

[4.1] Type of device (AP, AT, UXO etc.)	[4.2] Model	[4.3] Quantity	[4.4] Anti-lift fitted	[4.5] Booby trapped
			☐ Yes ☐ No	☐ Yes ☐ No
			☐ Yes ☐ No	☐ Yes ☐ No
			☐ Yes ☐ No	☐ Yes ☐ No
			☐ Yes ☐ No	☐ Yes ☐ No

IMSMA **Completion Survey**

Locator code: .../.../.../...

[5] Quantity and type of marked devices:

[5.1] Type of device (AP, AT, UXO etc.)	[5.2] Model	[5.3] Quantity	[5.4] Anti-lift fitted	[5.5] Booby trapped
			☐ Yes ☐ No	☐ Yes ☐ No
			☐ Yes ☐ No	☐ Yes ☐ No
			☐ Yes ☐ No	☐ Yes ☐ No
			☐ Yes ☐ No	☐ Yes ☐ No

[5] Status History

Status	Valid from	Status user	Reason	Comment

IMSMA **Completion Survey**

Locator code: .../.../.../...

[6] Declaration

Declaration by senior representative of the organisation that carried out the work.		**Declaration by responsible authority.**	
I declare that the area described in this document has been cleared in accordance with UN MAC Technical Guidelines, and that, to the best of my knowledge and belief, it is free of landmines and unexploded ordnance.		1) This declaration of clearance is accepted. 2) The area described in this report is accepted as clear of landmines and unexploded ordnance.	
[6.1]Name:		[6.4]Name:	
[6.2]Position:		[6.5]Organisation:	
[6.3]Date:		[6.6]Date:	
Signature:		Signature:	
This document has been distributed as shown below			
[6.7] **a.** Original to:		[6.9] **c.** Copy to:	
[6.8] **b.** Copy to:		[6.10] **d.** Copy to:	

[7]Attach explanatory map(s) and/or sketch(es) (required):
(Show on sketch type and location of cleared/marked area marking)

APPENDIX C

STATISTICAL ANALYSIS OF LANDMINE FATALITIES

Table C.A. Percentage share of killed and injured and share of killed by gender, 1970–2003.

	Number of victims			Share of victims		Share of killed	
Accident Time	Victims	Killed	Injured	Killed	Injured	Male	Female
1970	156	124	32	79.50	20.50	87.80	12.20
1971	6	5	1	83.30	16.70	83.30	16.70
1972	27	22	5	81.50	18.50	88.90	11.10
1973	21	16	5	76.20	23.80	95.20	4.80
1974	213	161	52	75.60	24.40	91.10	8.90
1975	113	90	23	79.60	20.40	93.80	6.20
1976	63	52	11	82.50	17.50	93.70	6.30
1977	49	42	7	85.70	14.30	98.00	2.00
1978	45	26	19	57.80	42.20	93.30	6.70
1979	53	46	7	86.80	13.20	90.60	9.40
1980	72	52	20	72.20	27.80	90.30	9.70
1981	63	55	12	82.10	17.90	91.00	9.00
1982	106	91	15	85.80	14.20	93.40	6.60
1983	125	109	16	87.20	12.80	87.20	12.80
1984	154	135	19	87.70	12.30	87.00	13.00
1985	207	160	47	77.30	22.70	90.80	9.20
1986	316	261	55	82.60	17.40	90.20	9.80
1987	374	316	58	84.50	15.50	91.40	8.60
1988	416	342	74	82.20	17.80	89.70	10.30
1989	209	174	35	83.30	16.70	92.30	7.70
1990	169	139	30	82.20	17.80	90.50	9.50
1991	2,366	1,904	462	80.50	19.50	87.30	12.70
1992	1,306	1,120	186	85.80	14.20	92.60	7.40
1993	986	859	127	87.10	12.90	93.20	6.80
1994	979	830	149	84.80	15.20	92.20	7.80
1995	945	814	131	86.10	13.90	91.60	8.40

1996	935	766	169	81.90	18.10	91.00	9.00
1997	792	665	127	84.00	16.00	93.70	6.30
1998	633	533	100	84.20	15.80	95.10	4.90
1999	464	359	105	77.40	22.60	92.50	7.50
2000	102	76	26	74.50	25.50	90.20	9.80
2001	138	135	3	97.80	2.20	94.20	5.80
2002	180	94	86	52.20	47.80	91.10	8.90
2003	87	10	77	11.50	88.50	98.90	1.10

Table C.B. Percentage share of killed among occupation groups

Accident Time	Killed	Pesh-marga	Child	Unemployed	Shepherd	Farmer	Driver	Others
1970	79.50	16.70	20.50	6.40	20.50	34.00	1.90	0.00
1971	83.30	16.70	0.00	16.70	50.00	16.70	0.00	0.00
1972	81.50	11.10	22.20	7.40	3.70	51.90	3.70	0.00
1973	76.20	0.00	19.00	4.80	14.30	57.10	0.00	4.80
1974	75.60	29.10	9.90	6.60	11.70	39.00	3.80	0.00
1975	79.60	14.20	12.40	5.30	18.60	44.20	5.30	0.00
1976	82.50	1.60	14.30	6.30	28.60	46.00	3.20	0.00
1977	85.70	8.20	22.40	4.10	18.40	42.90	4.10	0.00
1978	57.80	15.60	40.00	6.70	11.10	24.40	2.20	0.00
1979	86.80	17.00	18.90	5.70	35.80	22.60	0.00	0.00
1980	72.20	20.80	25.00	5.60	13.90	31.90	2.80	0.00
1981	82.10	20.90	19.40	3.00	11.90	43.30	1.50	0.00
1982	85.80	30.20	15.10	4.70	15.10	25.50	9.40	0.00
1983	87.20	15.20	29.60	12.80	8.80	27.20	5.60	0.80
1984	87.70	14.30	29.20	9.70	11.70	29.90	5.20	0.00
1985	77.30	21.30	17.90	7.70	17.40	31.40	4.30	0.00
1986	82.60	25.00	13.90	9.20	13.30	35.40	3.20	0.00
1987	84.50	20.30	23.50	7.50	9.60	35.00	4.00	0.00
1988	82.20	20.20	17.10	7.70	10.60	38.20	6.30	0.00
1989	83.30	16.70	23.40	6.20	15.80	32.10	5.70	0.00
1990	82.20	12.40	20.70	10.10	18.30	32.50	5.90	0.00
1991	80.50	8.90	34.80	10.20	5.00	37.30	3.70	0.10
1992	85.80	8.70	23.50	8.00	12.90	43.80	3.10	0.10
1993	87.10	6.80	20.40	6.60	15.00	47.30	3.80	0.20
1994	84.80	10.80	20.70	6.60	13.50	45.60	2.60	0.20
1995	86.10	12.50	21.70	7.60	12.20	43.40	2.60	0.00
1996	81.90	11.70	27.00	6.80	14.10	36.40	4.00	0.10
1997	84.00	17.70	19.70	6.70	15.20	37.90	2.90	0.00
1998	84.20	21.60	19.30	5.40	15.30	35.50	2.80	0.00
1999	77.40	19.20	25.00	6.30	22.60	22.40	4.50	0.00

2000	74.50	7.80	35.30	5.90	17.60	26.50	6.90	0.00
2001	97.80	4.30	27.50	4.30	23.90	34.10	5.10	0.70
2002	52.20	20.00	20.00	10.00	20.60	25.00	3.30	1.10
2003	11.50	4.60	13.80	4.60	14.90	43.70	12.60	5.70

Table C.C. Percentage share of killed among activity at time of accident

Accident Time	Killed	Walk	Touch	Collect	Fight	Fire	Others
1970	79.50	32.10	32.10	22.40	3.20	1.30	9.00
1971	83.30	16.70	50.00	33.30	0.00	0.00	0.00
1972	81.50	18.50	55.60	18.50	0.00	7.40	0.00
1973	76.20	14.30	61.90	14.30	0.00	0.00	9.50
1974	75.60	34.30	25.40	21.60	6.10	0.00	12.70
1975	79.60	37.20	21.20	27.40	3.50	1.80	8.80
1976	82.50	17.50	55.60	22.20	1.60	0.00	3.20
1977	85.70	26.50	24.50	32.70	0.00	2.00	14.30
1978	57.80	44.40	28.90	22.20	0.00	0.00	4.40
1979	86.80	34.00	20.80	45.30	0.00	0.00	0.00
1980	72.20	31.90	26.40	20.80	1.40	4.20	15.30
1981	82.10	43.30	25.40	11.90	4.50	1.50	13.40
1982	85.80	28.30	30.20	24.50	6.60	0.00	10.40
1983	87.20	28.80	31.20	21.60	3.20	4.00	11.20
1984	87.70	34.40	29.90	18.80	2.60	1.90	12.30
1985	77.30	27.10	28.00	24.60	4.30	0.50	15.50
1986	82.60	35.40	25.60	20.30	4.70	1.30	12.70
1987	84.50	37.70	24.90	20.60	2.70	2.70	11.50
1988	82.20	45.00	23.80	15.10	2.20	1.70	12.30
1989	83.30	29.70	34.40	27.30	1.90	1.00	5.70
1990	82.20	30.20	21.90	37.30	0.00	0.60	10.10
1991	80.50	37.00	35.00	16.10	1.10	1.60	9.30
1992	85.80	26.80	34.60	29.60	1.80	2.10	5.00
1993	87.10	23.90	33.30	34.10	1.60	1.60	5.50
1994	84.80	26.00	32.60	33.60	0.80	1.00	5.90
1995	86.10	27.10	30.10	31.60	1.50	1.50	8.30
1996	81.90	22.10	35.90	29.90	1.80	2.00	8.10
1997	84.00	28.20	31.80	27.30	3.50	3.20	6.10
1998	84.20	21.60	31.90	26.50	6.50	2.20	11.20
1999	77.40	20.70	36.90	27.40	3.90	1.70	9.50
2000	74.50	20.60	45.10	19.60	0.00	2.90	11.80
2001	97.80	9.40	39.90	25.40	0.70	9.40	15.20
2002	52.20	12.20	31.10	35.00	7.80	2.80	11.10
2003	11.50	0.00	8.00	88.50	2.30	1.10	0.00

Table C.D. Population, victims and victims killed and injured per 1,000 inhabitants by districts.

	Population			Fatalities			Fatalities per 1000	
District	Male	Female	Total	Injured	Killed	Total	Injured	Killed
Akre	102,794	97,885	200,679	20	70	90	0.10	0.35
Amedi	51,929	49,979	101,908	46	329	375	0.45	3.23
Chamchamal	65,659	67,857	133,516	26	321	347	0.19	2.40
Choman	18,533	18,185	36,718	175	769	944	4.77	20.94
Darbandikhan	21,672	21,897	43,569	12	443	455	0.28	10.17
Dashti Hawler	65,835	65,298	131,133	86	100	186	0.66	0.76
Duhok	142,330	136,977	279,307	57	120	177	0.20	0.43
Dukan	29,089	30,045	59,134	212	197	409	3.59	3.33
Erbil	614,389	604,523	1,218,912	117	324	441	0.10	0.27
Halabja	50,439	51,737	102,176	38	800	838	0.37	7.83
Kalar	78,756	78,144	156,900	4	488	492	0.03	3.11
Khabat	32,873	32,103	64,976	1	0	1	0.02	0.00
Khanaqin	..	..	..	5	218	223	..	..
Kiffri	18,418	18,585	37,003	8	161	169	0.22	4.35
Koye	58,728	56,639	115,367	11	300	311	0.10	2.60
Makhmur	..	..	..	1	0	1	..	..
Mergasur	31,286	32,231	63,517	40	201	241	0.63	3.16
Penjwin	24,396	24,358	48,754	91	1,010	1,101	1.87	20.72
Pishdar	59,681	60,257	119,938	380	620	1,000	3.17	5.17
Rania	105,367	108,538	213,905	188	215	403	0.88	1.01
Semel	55,414	53,129	108,543	9	31	40	0.08	0.29
Shaqlawa	76,502	74,456	150,958	131	288	419	0.87	1.91
Sharazur	74,869	75,579	150,448	5	173	178	0.03	1.15
Sharbazher	10,570	10,526	21,096	171	1,241	1,412	8.11	58.83
Shekhan	36,343	35,404	71,747	6	46	52	0.08	0.64
Soran	120,920	117,185	238,105	242	870	1,112	1.02	3.65
Suleimaniya	408,907	422,588	831,495	167	934	1,101	0.20	1.12
Zakho	107,546	103386	210,932	41	305	346	0.19	1.45

Fig. C.A1. Number of killed and injured, 1970–2003.

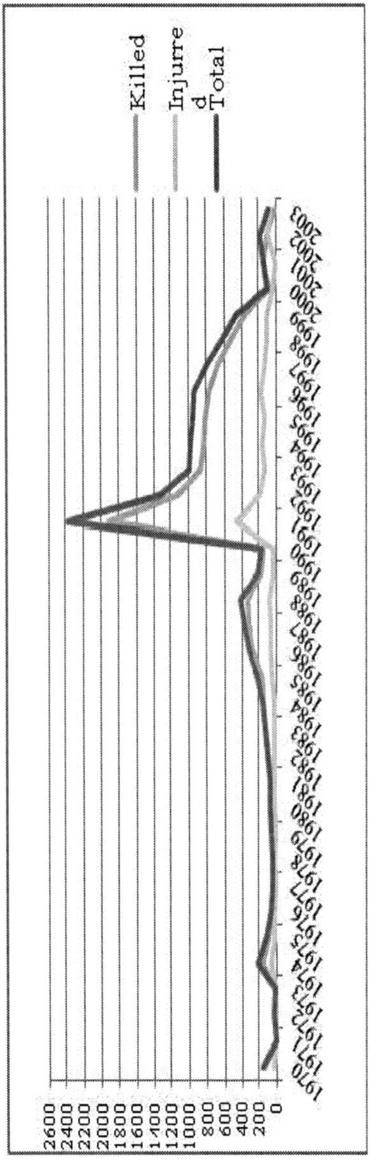

Fig. C.A2. Number of killed (male) and injured (female) by gender, 1970–2003.

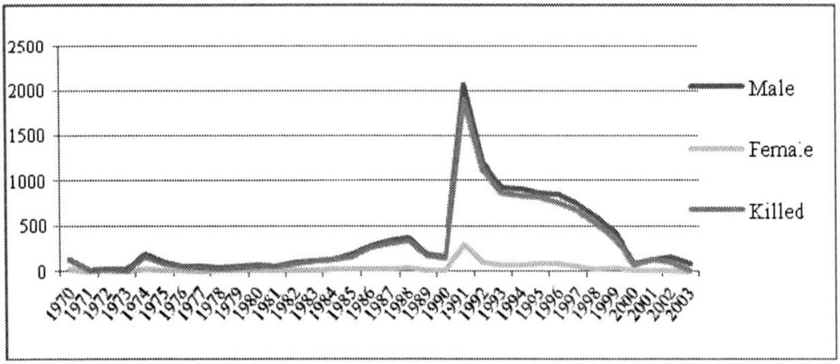

Fig. C.A3. Share of killed and injured, 1970–2003.

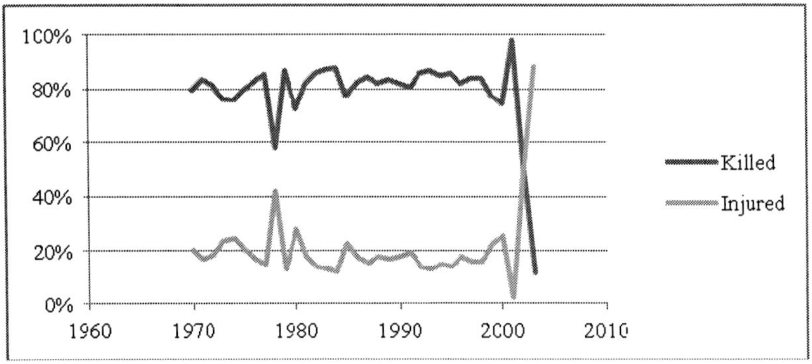

Fig. C.B1. Share of killed among different occupation groups (peshmarga, child, unemployed, shepherd, farmer, driver, others), 1970–2003.

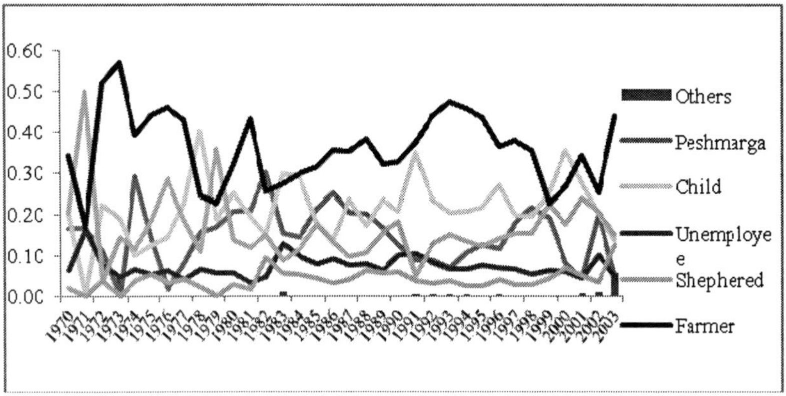

Fig. C.B2. Number of killed among different occupation groups (peshmarga, child, unemployed, shepherd, farmer, driver, others), 1970–2003.

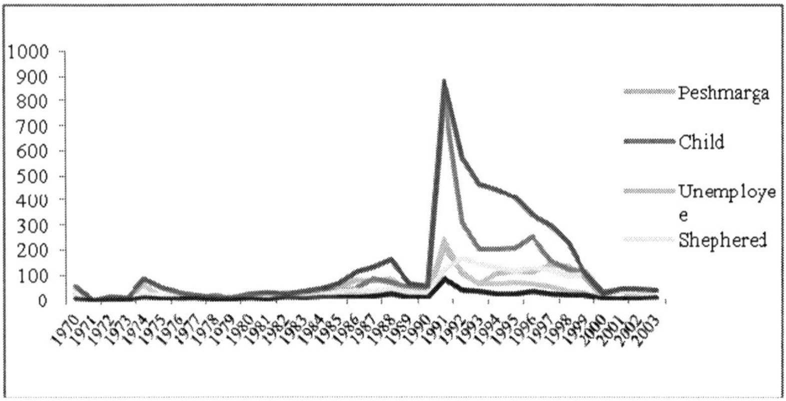

Fig. C.C1. Share of killed among different activity groups (walking, touching, collecting, fighting, fire, other activities) at the time of accident, 1970-2003.

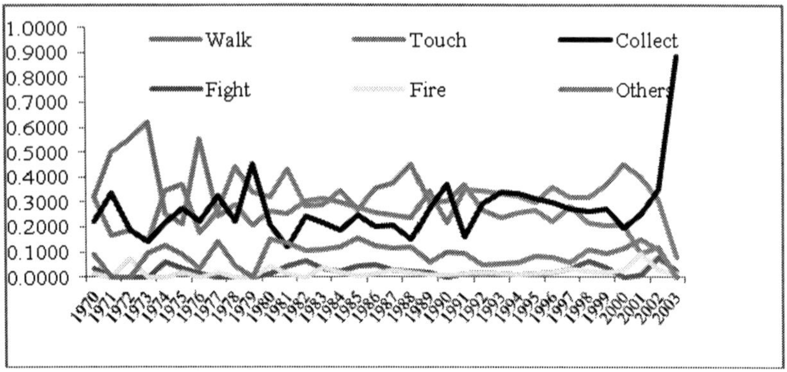

Fig. C.C2. Number of killed among different activity groups (walking, touching, collecting, fighting, fire, others) at the time of accident, 1970–2003.

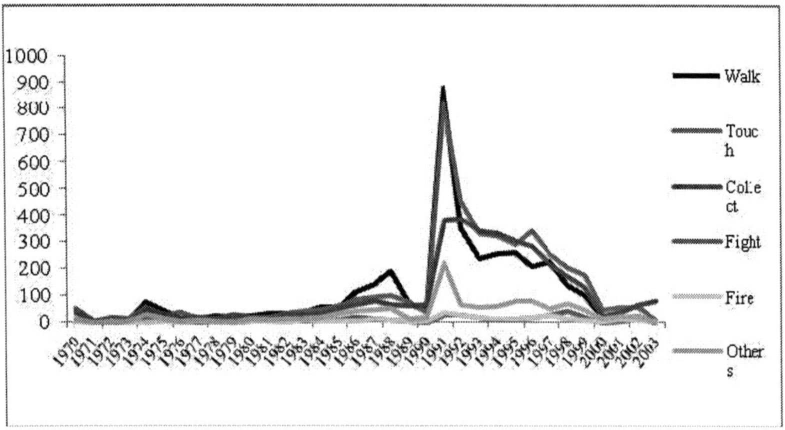

Fig. C.D1. Share of killed aware of danger (danger) and participating in mine awareness programs (mre), 1970–2003.

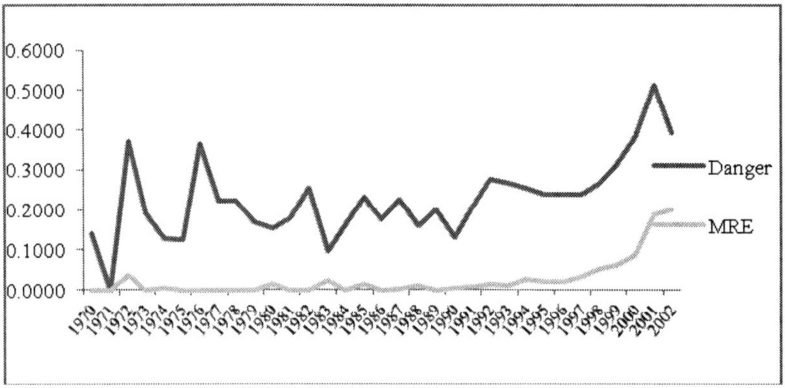

Fig. C.D2. Number of killed aware of danger (danger) and participating in mine awareness programs (mre), 1970–2003.

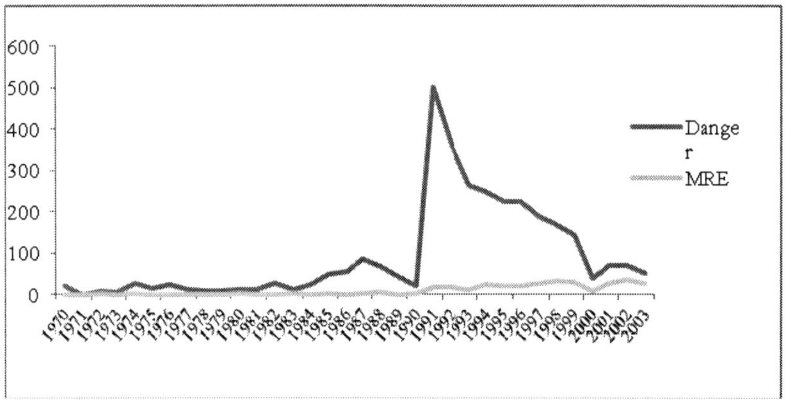

REFERENCES

A Study of the Role of Survey in Mine Action (2006), GICHD, Geneva Available at http://www.gichd.ch/fileadmin/pdf/publications/Survey_in_MA_March2006.pdf

Al-Bazzaz, P.H. (2010), "Kidney Transplantation in Erbil, Iraq: A Single-center Experience", Saudi Journal of Kidney Diseases and Transplantation, 21(2), 359-62.

Alkire, S. (2003), "A Conceptual Framework for Human Security", Centre for Research on Inequality, Human Security, and Ethnicity (CRISE), Working Paper 2, London: University of Oxford.

Andersson, N., da Sousa, C.P. and Paredes, S. (1995), "Social Costs of Landmines in Four Countries: Afghanistan, Bosnia, Cambodia, and Mozambique", British Medical Journal, 311, 718–721.

Anti-Personal Landmines, Friend or Foe, (1997), "A Study of the Military Use and Effectiveness of Anti-personnel Mines", ICRC 1997, available at http://www.icrc.org.

Ascherio, A., Biellik, R., Epstein, A., Snetro, G., Gloyd, S., Ayotte, B. and Epstein, P.R. (1995), "Deaths and Injuries Caused by Landmines in Mozambique", Lancet, 346, 721-725.

Askin S. and Goose, S. (1994), "The Market for Antipersonnel Landmines - A Global Survey", Jane's Intelligence Review.

Bajpai, K. (2000), "Human Security: Concept and Measurement", Kroc Institute Occasional Paper, 19(1), 103-125.

Baltagi, B.H. (2008), "Econometric Analysis of Panel Data", Fourth Edition, John Wiley and Sons, Ltd.

Beach, H. (2001), "Cluster Bombs: The Case for New Controls", International Security Information Service, Briefing Paper No 25.

Benedinelli, C. (2009), "Effects of Landmines and Unexploded Ordnance on the Pediatric Population and Comparison with Adults in Rural Cambodia", World Journal of Surgery, 33(5), 1070-1074.

Berhe, A.A. (2007), "The Contribution of Landmines to Land Degradation", Land Degradation and Development, 18(1), 1-15.

Berthiaume, A. (2003), "Gender and Landmines", Resource prepared for Youth Mine Action Ambassador Program, Ottawa, Canada: International Campaign to Ban Landmines.

Bier, G.L. (2003), "The Economic Impact of Landmines on Developing Countries", International Journal of Social Economics, 30(5), 651-662.
Bottomly, R. (2003), "Crossing the Divide: Landmines, Villagers and Organisations", Oslo, Norway: International Peace Research Institute.
Brealey, R.A. and Stewart C.M. (1991), "Principles of Corporate Finance", 4th ed. McGraw-Hill.
Bugnion, F. (2002), "Just War. Wars of Aggression and International Humanitarian Law", International Committee of the Red Cross 84(847), 15-23.
Byrd, A.W. and Gildestad, B. (2001), "The Socio-Economic Impact of Mine Action in Afghanistan. A Cost-Benefit Analysis", Geneva International Center for Humanitarian Demining GICHD, Available at http://www.gichd.org/fileadmin/pdf/ma_development/database/Study-SocioEco-Impact-of-Mines-in-Afghanistan-Gilestad-2001.pdf
Carnahan, B.M. (1984), "The Law of Land Mine Warfare: Protocol II to the United Nations Convention on Certain Conventional Weapons", Military Law Review 105, 73-75.
Champion, H.J.R., Sacco, W.J. and Copes, W.S. (1989), "A Revision of the Trauma Score", Journal of Trauma, 29, 623-629.
Coupland, R.M. (1996), "The Effect of Weapons: Defining Superfluous Injury and Unnecessary Suffering", A1 Medicine and Global Survival, 3, 1-7.
Davis, P. and N. Duniop (1994), "War of Mines, Cambodia, Landmines and the Improvements of Nation", Pluto press, London.
De Mulinen, F. (1989), "Hand Book of the law of War for armed forces", International Committee of the Red Cross, Geneva 1989.
Deaton, A. (1985), "Panel Data from Time Series of Cross Sections", Journal of Econometrics, 30, 109-126.
Elie, F.A. (201), "Human Security: Consequence of and Incentive for National and International Security", Journal of Defense Resources Management, 2(2), 105-110.
Engle, E.A. (2009), "The History of the General Principle of Proportionality: An Overview", Willamette Journal of International Law and Dispute Resolution, Forthcoming. Available at SSRN: http://ssrn.com/abstract=1431179 .
Gangwar, A. (2003), "Impact of War and Landmines on Environment", Centre for Environment Education, Available at http://www.mtnforum.org/sites/default/files/pub/1409.pdf.
Geneva Intention Centre for Humanitarian Demining, GICHD, www.gichd.org

Ghali, B. (1994), “The Land Mine Crisis: A Humanitarian Disaster”, Foreign Affairs, 73(5), 8-13.
Goose S. (1995), "The Economics of Landmines", article for UNIDIR Newsletter, published in early 1995, citing US Army Foreign Science and Technology Center, US Defense Intelligence Agency, et al, Landmine Warfare - Mines and Engineer Munitions in Southern Africa, May 1993, p. 15.
Grant, R. (2003), "In Determining Military Necessity and Proportionality, The Commander's Judgment is more Critical than Ever, in search of Lawful Targets", Air Force Magazine, February 2003, 38-44.
Greene, W. (2008), “Econometric Analysis”, Prentice Hall, 6th Edition.
Hampson, F. (2002), Madness in the multitude: human security and world disorder, Oxford University Press.
Harpviken, K.B. (2003), “The Future of Humanitarian Mine Action: Introduction”, Third World Quarterly, 24(5), 777-780.
Harris, G. (2000), “The Economics of Landmine Clearance: Case Study of Cambodia”, Journal of International Development, 12(2), 219-225.
Heshmati, A. and Kumbhakar, S.C. (1997), “Estimation of Technical Efficiency in Swedish Crop Farms: A Pseudo Panel Data Approach”, Journal of Agricultural Economics, 48(1), 22-37.
Human Development Report (1994), Published for the United Nations Development Program (UNDP), Oxford University Press.
Human Rights Watch (2002), “Fatally flawed: Cluster Bombs and their use by the United States in Afghanistan, 14(7) (G), p.26.
Husum, H., Resell, K., Vorren, G., Van Heng, Y., Murad, M., Gilbert, M. and Wisborg, T. (2002), “Chronic Pain in Landmine Accident Survivors in Cambodia and Kurdistan”, Social Science and Medicine, 55, 1813-1816.
ICC (1997), Report of the ICRC Mine Awareness Programme. Geneva: International Committee of the Red Cross; 1997.
—. (2002), Anti-personnel landmines around the world: ICRC mine action and the Ottawa Conference. Geneva: International Committee of the Red Cross; 2002. ICRC Fact Sheet 28 Nov.
IMAS (2007), “Guide for the establishment of a mine action programme”, IMAS 02.10, First Edition, 01 August 2007.
IMAS Mine Risk Education Best Practice Guide Book1 (2005), “An Introduction to Mine Risk Education”, UNICEF.
International Committee of the Red Cross ICRC (1999), “Daily Bulletin of the 27th International Conference of the Red Cross and Red Crescent”.

International Save the Children Alliance (2000), "Child Landmine Survivors: An Inclusive Approach to Policy and Practice", Working Paper Number 2.

Jahanlu, H.R., Husum, H. and Wisborg, T. (2002), "Prehospital Mortality in Landmine Accidents in Iran", Prehospital Disaster Medicine, 17(2), 107-109.

Kakar, F. (1995), "Direct and Indirect Consequences of Landmines on Public Health", Geneva: World Health Organization.

Kakar F., F. Bassani, C.J. Romer and S.W.A. Gunn (1996), The consequences of land mines on public health. Prehospital and Disaster Medicine, 11(1), 2-10.

Kievelitz, U., T., Schaef, M. Leonhardt, H.H.,and Vorwerk, S. (2004), "Practical Guide to Multilateral Needs Assessments in Post-Conflict Situations, A Joint UNDG, UNDP and World Bank Guide, prepared by GTZ with the support of BMZ", Conflict Prevention & Reconstruction Paper No. 15, The World Bank, Washington DC.

King, C. (1996), "Janes Mines and Mine Clearance Techniques", in King, C. (Ed) Coulsdon (163 Brighton Rd, Coulsdon, Surrey CR5 2NH, UK) Jane's Information Group.

King, G., Christopher, M. (2002), "Rethinking Human Security", Political Science Quarterly, 116(4), 585-610.

Krug, E. G., Ikeda, M.R., Qualls, L.M., Anderson, A. M., Rosenberg, L., M., and Jackson, J.R. (1998), "Preventing Landmine Related Injuries and Disability", Journal of the American Medical Association, 280(5), 465-466.

Landmine Impact Survey Report (1999-2003), Implemented by UNOPS MAP in Northern Iraq under the Oil for Food Program.

Landmine Impact Survey-The Republic of Iraq (2004-2006), Implemented by the Information Management & Mine Action Programs, VVAF Iraq.

Matthew, R., A, McDonald, B., and Rutherford, K.R. (2004), "Landmines and Human Security: International Politics and War's Hidden Legacy", State University of New York Press.

Meyrowitz, H. (1994),"The Principle of Superfluous Injury or Unnecessary Suffering- From Declaration of St Petersburg of 1868 to Additional Protocol 1 of 1977", International Review of the Red Cross, 34(299), 98-122.

Mine Action and effective coordination (2003),"The United Nations Policy, Sectorial Policy: The Scope of Action of Mine Action Centers and Organizations in Victim Assistance", UNMAS.

Mine Action Information Center (2002), "The Landmine Action Smart Book", First Edition, Harrisonburg, VA.

Mine Action in the three Northern governorates of Iraq (2003), Situation Report, UN Office for Project Services, available at http://reliefweb.int/node/104651

Mine Advisory Group Iraq, MAG Iraq, www.magiraq,org

Mine Advisory Group International, MAG International, www.maginternational.org

Morin, A. (1998), "Demining and the Environment: A primer", Journal of Mine Action, 11(2). Available at http://maic.jmu.edu/journal/11.2/feature/morin/morin.htm

Newman, R.D. and Mercer, M.A. (2000), "Environmental Health Consequences of Landmines", International Journal of Occupational Environment and Health, 6(3), 243-248.

Ouarda, M. (2008), "Landmines and Poverty IV: Evidence from Mozambique", Peace Economics, Peace Science and Public Policy, 4(1), 125-153.

Paterson T. (2001), "Commentary on the Economics of Landmine Clearance: Case Study of Cambodia", Journal of International Development, 13, 629-634.

Pictet, J. (1966), "The Principle of International Humanitarian Law", Halaman 10.

Quesada, J. (1998), "Suffering Child: An Embodiment of War and its Aftermath in Post-Sandinista Nicaragua", Medical Anthropology Quarterly, 12, 51-73.

Roberts, S. and Williams, J. (1995), "After the Guns Fall Silent. The Enduring Legacy of Landmines", Washington, DC: Vietnam Veterans of America Foundation.

Sandoz, Y. (1981), "A New Step Forward in International Law-Prohibitions and Restrictions on the Use of Certain Conventional Weapons", International Review of the Red Cross, 220, 41-56.

Shabila, N.P., Taha, H.I. and Al-Hadithi, T.S. (2010), "Landmine Injuries at the Emergency Management Center in Erbil, Iraq", Conflict and Health, 4, 15, 1-6.

Steven, B, and Tillery, G. (1996), "Cost and Effectiveness Modeling for Demining Operations", Technical Report No. 38.

Sunga, L. S. (2009), "The Concept of Human Security: Does it Add Anything of Value to International Legal Theory or Practice?", in Power and Justice in International Relations Interdisciplinary Approaches to Global Challenges Power and Justice in International Relations (Edited by Marie-Luisa Frick and Andreas Oberprantacher), Ashgate Publishers.

Themnér, L. and P. Wallensteen (2011), "Armed Conflict, 1946-2010", The Journal of Peace Research, 48(4), 525-536.
UNDP Fact Sheet (2009), "Bureau for crisis prevention and recovery", Mine Action statistics.
UNICEF (2003), "Impact of Landmines on Children", East Asia and Pacific Regional Office.
United Nations Developing Program (2009), "Overview of Landmines and Explosive Remnants of War in Iraq", UNDP.
United Nations fund for women (2003), "Issue Brief on Landmines", available at www.womenwarpeace.org.
United Nations Mine Action Services (UNMAS), International Mine Action Standards (IMAS), New York: UNMAS, 2001, available at: http://www.mineactionstandards.org/imas.htm.
United Nations Iraq, www.uniraq.org
Verbeek, M. (1992), "Pseudo Panel Data", in Matyas L. and Sevestre P. (eds), The Econometrics of Panel Data, 303-314, Kluwer Academic, Amsterdam.
Verbeek, M., Nijman, T. (1992), "Can Cohort Data be Treated as Genuine Panel Data?", Empirical Economics, 17, 9-23.
Walsh, N.E. and W.S. Walsh (2003), "Rehabilitation of Landmine Victims- The Ultimate Challenge", Bulletin of the World Health Organization, 81(9), 665-670.
Wheeler, D. and Tiefelsdorf, M. (2005), "Multicollinearity and Correlation among Local Regression Coefficients in Geographically Weighted Regression", Journal of Geographic System, 7(3), 161-187.
Wooldridge, J.M. (2006), "Introductory Econometrics", 4th Edition, Michigan State University.

INDEX OF AUTHORS

INDEX OF SUBJECTS